ATOMIC IMPACT

Systems for Transformative Productivity

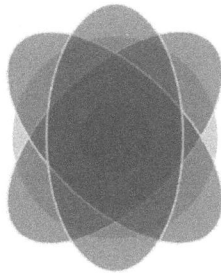

Dr. Gilbert A. Guzman

Gilbert A. Guzman
Fort Worth, Texas

Publisher's Note. *Atomic Impact* is a work of nonfiction grounded in the author's professional experience, research, and personal insights. While real events, organizations, and individuals are referenced or anonymized to protect confidentiality, any illustrative scenarios or case examples are presented for educational purposes and may include composites or fictionalized elements to support clarity. The views expressed are those of the author and do not necessarily reflect the opinions or positions of any affiliated organizations.

Book Layout © 2017 BookDesignTemplates.com

Atomic Impact/ Dr. Gilbert A. Guzman. , 1st ed.
ISBN 979–8–9991043–0–4

Dedication

To my lovely wife and children - the roots of my strength and the reason I grow. With your love, I've learned that true growth isn't just about reaching higher, but becoming better... together.
—Dr. Gilbert A. Guzman

Contents

Why Small Actions Create Massive Impact

Atomic Impact is a revolutionary approach to productivity that focuses on the strategic role of systems, design thinking, and technology in driving organizational success. While acknowledging the importance of small, deliberate actions, this book shifts the focus to how these actions can be amplified within structured systems to meet operational demands effectively.

Think of the impact through the lens of an atomic bomb. Discovering such power through iteration after iteration, theory upon theory, to later become one of the most dangerous, yet powerful weapons of its time, the atomic bomb began as merely an idea. How could something so small become so powerful?

Built upon key frameworks from various scholars and thought leaders such as Dr. W. Edwards Deming, Dr. Jay Forrester, and many others, this book serves as a guide for students, professionals, and organizational leaders looking to achieve transformational productivity through systemic thinking by acknowledging the atomic impacts of the work our teams do every day.

At its core, this book aims to redefine productivity by integrating foundational actions into broader systems. By leveraging design principles and operational strategies, readers will discover how to create a culture of sustainable efficiency and innovation by ensuring the small things count.

But let me be honest with you: this book was born from frustration. After finishing my doctorate during the pandemic, watching The Great Resignation unfold in real time, and seeing organization after organization fail despite having all the "right" strategies, I knew something fundamental was broken in how we think about productivity.

This isn't a book about working harder. It's about understanding that in every system, there are atomic actions - small, specific interventions that can trigger chain reactions of productivity. Just as splitting a single atom can release enormous energy, the right action at the right point in your system can transform everything.

Now you understand the promise of atomic impact. But promises need foundations. That foundation is systems thinking. Atomic actions without systems are just good intentions.

PART I:

Understanding Systems

CHAPTER 1

Systems as the Foundation for Productivity

Productivity doesn't occur in a vacuum.

This chapter explores how systems, both formal and informal, create the backbone for individual and organizational performance.

Understanding System Design

When I first decided to write this book, I had just finished my doctorate (DBA). I was fixated with this world of systems and system design. At this point, writing for nearly four years, I had the opportunity to expand my knowledge into the topic of system design and systems thinking. I remember citing Jay Forrester in my dissertation, but I was not familiar with his work on System Dynamics as a broad framework since my dissertation was focused on W. Edwards Deming's System of Profound Knowledge and Fred Davis's Technology Acceptance Model. Knowing that Deming was focused on systems and the interdependence of systems, I knew there had to be more to discover. Perhaps I was at a point where my thoughts and day-to-day job duties always involved some sort of problem or challenge and the need to solve it.

During my doctoral coursework, specifically my course called Fundamentals of Productivity, my mind always wandered to how everything I touched at work was part of a system that seemed to be out of sync. Some component or subsystem within the larger system was not firing as intended. By understanding that systems are composed of interrelated components that work together to achieve a purpose, effective systems are intentionally designed to align resources, processes, and outcomes. What happens when these resources, processes, and outcomes are not truly aligned? Can we still call "it" a system?

Taking on research, falling in deeply with the concept of productivity, and thinking about how systems are interconnected, I knew there had to be something that would drive me toward a novel theory or framework. As I began developing computational models of organizational behavior, I realized I needed to build upon existing frameworks.

It's important to note from the outset that systems don't function in isolation from the people within them. For any system to work effectively, there must be psychological safety - a belief that one will not be punished or humiliated for speaking up with ideas, questions, concerns, or mistakes. This concept, which we'll explore in depth in Chapter 15, is the invisible infrastructure that enables atomic actions to propagate through systems. Without it, even the best-designed systems fail because people won't take the small risks necessary for innovation and improvement.

Deming's System of Profound Knowledge

As Deming developed his System of Profound Knowledge (SoPK), he references that people and processes together define a system whereas an understanding of this relationship is necessary for any system to work as it is intended. With the four core values or principles of the SoPK, it is critical to understand that it too is a system of cross-connected components in which decision-makers, the leaders and managers of our organizations, make decisions that are comprised of complexity, primarily in the realm of human variability, irrationality, and unpredictability.

The appreciation for a system: Managers and leaders must ensure that everyone understands the aim of the system they operate in or are a part of.

Knowledge about variation: A constant, things will always vary. We need to understand and know if it is a common or special variation.

The theory of knowledge: A plan is simply a prediction based on previous system results. Knowledge of these results help us predict outcomes and make decisions.

Knowledge of psychology: People are different, and as components of a system (often), they are not to be ranked based on performance.

The SoPK in its most simple form is knowing the relationships, roles, and variability that exists in a system. Manage this and you at least have a chance to succeed.

The System Dynamics Revolution

Since this chapter is largely about system design, let's talk system design. As my initial research was fixated on Deming's SoPK, glancing into the world of system dynamics and meeting wonderful people from the Massachusetts Institute of Technology (MIT), Jay Forrester's home for nearly fifty years, I knew there was a place where I can build upon my knowledge of systems thinking beyond Deming's SoPK.

Aside from reading a little on Forrester, I first came across a YouTube lecture and introduction to System Dynamics by Professor John Sterman which then led me to a graduate lecture by James Paine (now Dr. James Paine). Both lectures introduced system dynamics with a touch on systems thinking. Listening to both lectures prompted me to purchase Professor Sterman's textbook titled *Business Dynamics: Systems Thinking and Modeling for a Complex World*. Professor Sterman warned all his students attending the lecture that it is a large book, but it will cure their insomnia and with two, they can cancel their gym memberships (if they had one).

Both Sterman and Paine drove deep into the concepts of data and the power of data that aligns our challenges with data. Forrester mentioned this in a 1988 lecture where he said that as leaders and managers, our greatest strength and weakness are that we are "overwhelmed with information, and perhaps one of the most serious pollutions that we face is information pollution...too much information." While we know we have the power of data, analytics, data science, business intelligence, etc. today, this notion rang true in 1988! We didn't have the World Wide Web yet (www, remember having to type that in the front of your url?), and it rang true!

Mental Models and Open Loop Thinking

With information pollution, we often think that our "mental models", the way we think, are what drive the success of the systems we have, to achieve a goal. This leads to the concept or framework of open loop thinking. This is what drives our day-to-day (at times). We have goals, all of which are often implemented based on problems identified by gathering data and or information, maybe "polluting" our thought process, then we make decisions based on this information, and ultimately expect those decisions to help us reach the goal.

This is largely controversial in the realm of systems thinking because as Dr. James Paine said, "No decision you make exists in isolation; no process that you affect exists in isolation." What does he mean by this? Well, take for example a goal of achieving a hiring target. This hiring target exists merely because you have a problem of turnover, which ultimately challenges your bottom line/profitability because you have contractual obligations to provide a service at a contracted rate without the need for overtime. This problem of turnover leads to overtime because you are contractually obligated to provide human capital. To ensure you meet your obligations, you fill it with an individual who will work up to eight hours of overtime.

The problem that we see in this process is the overtime, which has direct financial implications on your bottom line. The solution is to hire. The turnover, how many needs we fill with overtime, etc. are all data points and information that drive our decision(s). These decisions might be how many people we need to hire to ensure we fill all the needs without overtime. That is where the mental model comes in. We generate a decision to hire five more people than we terminate each week, and ultimately that will solve our problem.

However, systems thinking requires us to understand that our decisions are impacted by the decisions of others, and on the other side of that coin, our decisions ultimately impact others. Additionally, we happen to misunderstand the feedback loops that exist in a system. We must acknowledge that systems, whether physical or mental, are not linear in nature. Every action has a reaction; however, every reaction might lead to another action. Sometimes the outputs we expect in a system are generated by actions we did not intend. My simulations had to account for this reality: organizational actions create ripple effects that propagate through networks in ways that linear thinking cannot predict.

Why "Atomic"?

So, you are probably wondering why I titled the book *Atomic Impact*, right? Let's go way back , and no it isn't because of James Clear's *Atomic Habits* or Leo Black's *Atomic Productivity* (both great books, by the way). In high school, I had a physics teacher named Mrs. Karen Held. While obviously I did not follow the path of science, I have always been intrigued by science, physics, math, you name it. Mrs. Held can be thanked for some of this interest as she made the classroom experience fun, often tying in the concepts with day-to-day life or some sort of experiment.

As children and younger teens, my brothers and I were fascinated with the show *Modern Marvels*, *How It's Made*, *National Geographic*, etc. (One became an electrical engineer and the other a dexterous tinkerer.) Through this intriguing foundation, I have always seemed to align whatever I was doing with some sort of science or math. Take for example, today: I often look through the lens of data science as much as possible to generate narratives about performance, human behavior, you name it!

With a mind that tends to wander and search for correlations, I often found myself looking for movies or shows to help understand the world. What possibilities exist? What other perspectives lie beyond my own mental models or paradigms of understanding? I think of movies like *Interstellar* or *Limitless* that might have made me think of whether there was more beyond what is in front of me. Of course, finding live episodes of *Modern Marvels* and *How It's Made* on Prime Video is always a plus.

Fast forward to 2023–24 when I was afforded the time to sit down and watch Oppenheimer. It took me a few times to sit through and watch the movie since having a large family often took precedent over sitting through a three-hour movie of something they might not have understood or enjoyed. I finally watched the movie through the end and found myself watching it three or four more times, times that included my large family and my mother and father-in-law.

My wife and her family are from a small northern New Mexico town, a town that has been served by the very Los Alamos Labs depicted in Oppenheimer. My father-in-law worked in those labs for more than thirty-five years, and so too did his family, my mother-in-law's family, and almost the entire town. This movie helps explain why my in-laws and their families worked in those labs and why these labs have been a pinnacle of their town for more than eighty years, so I decided to understand more. While I knew about Robert J. Oppenheimer, I was not familiar with his work or those involved in the race to the degree that was portrayed in the movie. While Oppenheimer was the primary character, I needed to understand more about the men, the project, the science. This gave me the urge to rush over to Half Price Books and delve into the science section just to find anything on Quantum Physics, Quantum Mechanics, Quantum Theory or even the Atomic Bomb. During my search I came across a small book titled *Niels Bohr & Max Planck, Quantum Theory*. This little paperback book with 256 pages opened my mind a bit to what was not understood but was changed by the smallest of energy and action.

The Quantum Connection

As I read that book, rather slowly, to understand the concepts and theory, I remember holding a conversation with my Senior Vice President, Morgan P. over Texas Barbecue one evening. He had mentioned he has yet to see the movie Oppenheimer. See, Morgan, in my eyes, is an intellectual. He often listens to Malcolm Gladwell's *Revisionist History* Podcast, reads books, and has movie posters on his family home walls. I would have definitely expected he would have seen the movie by then. While he had not seen the movie, our conversation went down a rabbit hole that led him to recommending a book titled *The Making of The Atomic Bomb* by Richard Rhodes, a Pulitzer Prize winner in 1988. Morgan mentioned that the book was referred to him by his mother. This book, written in 1986, also softback with 896 pages, led me into more of what this world of quantum theory, atomic energy, the Manhattan Project, etc. were all about.

Through both of these books, beginning with Quantum Theory, I learned that the smallest of changes and considerations led to large impacts. While largely theoretical, much of the theories made sense. Take for example, Max Planck's constant (h) , we won't get into the actual formula, too much. This constant in its most basic form tells us that energy isn't continuous, and it works in the smallest levels, typically at an atomic level. Since this constant (h) references that energy isn't continuous, Planck teaches us that energy essentially comes in small packets or groups called *quanta*. This gives us the concept of quantized energy. This small calculation by Planck adding (h) to a formula helped him and others be able to track the frequency of the energy.

Learning this, my mind right away went to my passion for productivity and continuous improvement, leading me to write this book. Here lies the premise, the foundation, the epiphany for this book and the title, *Atomic Impact*. If energy is quantized and often seen in packets rather than a continuous, smooth flow, can productivity in the context of business considerations be evaluated with a constant such as (h)? If productivity is measured in its most basic form as Outputs/Inputs, can a constant (h) be added to the formula to help track the small, atomic contributions to productivity and performance? Can we measure what our laborers, workers, and "inputters" do at an atomic level that can mean more to an organization, process, or system than we account for? What is our smallest degree of measurement when it comes to the actions of humans and/or technology in a business?

The Division of Labor Reimagined

Adam Smith, in 1776, wrote in his book, *Wealth of Nations*, using his pin worker example that the division of labor is important due to the outputs. Considering specialized individuals and roles, outputs are far greater than having a single individual do the job(s) of many.

So, can we truly look at productivity in its most basic form and determine what the smallest input can be and maximize its contribution? When we think of productivity, we often think about the maximum effort that should ultimately create a maximum output. This effort should generate a contribution, another metric to account for.

Take for example a past life when I worked at Kroger, a leading US grocery chain. To measure productivity, the company implemented a workforce/labor metric that emphasizes the effectiveness of allocated labor (both variable and fixed) against sales volume over a week. This metric was tied to the effectiveness of the inputs based on a 15-minute rating where tasks, in the simplest form, were grouped by and measured in 15-minute increments. Visualize a bakery in your local grocery store. Often there are in-house produced items as well as prepackaged, par–baked, and frozen items that are sold in this department. Now, visualize a cake, one for your 20th anniversary at work. In the world of Kroger, the time it takes to decorate a cake will have a fixed and/or variable time to complete (input). Using forecasting from previous sales data, these times will be calculated based on how many cakes will more than likely need to be decorated, which ultimately generates the amount of labor necessary to generate the sale(s).

Through these parameters, we have the realm of efficiency. If productivity is outputs/inputs, efficiency, how well we use resources such as energy, time, money, people, etc., is critical to ensure the laborer (worker) meets the overall expectation. This, in an aggregated total that ultimately generates your headcount and is an important step because this tells us the "effectiveness" of the department, store, district, region/division, company. So, take into the inputs of a newly hired individual, such as a newly hired cake decorator. While these parameters are not explicitly explained or instructed to the cake decorator, the expectations are implied. Now, this newly hired cake decorator, while he/she will not be hired without some skillset and background in cake decorating, will not be 100 percent "effective" due to variables such as getting used to the department, tool location, thickness of the icing (yes, this varies sometimes), cake variations, etc.

The Manhattan Project as System Design

Now, back to a little more on the atomic consideration but through the lens of system design. The creation of the atomic bomb serves as a compelling illustration of system design in action. The Manhattan Project, a vast, multidisciplinary effort during World War II, brought together scientists, engineers, and military personnel to achieve a singular, highly complex objective. Key figures, such as J. Robert Oppenheimer, Enrico Fermi, Niels Bohr, and Max Planck exemplified the power of coordinated systems in managing resources, knowledge, and logistics.

Niels Bohr, building on Planck's quantum theory and contributing significantly to understanding atomic structures, used Planck's constant *(h)* to explain quantized energy levels in atoms. In an atom, energy is quantized, where only specific, discrete energy levels are allowed. These small, individual packets of energy, though seemingly insignificant, determine the atom's behavior and interactions. Similarly, in a workplace system, individual contributions of effort, no matter how small, act as the "energy packets" that drive the system forward. Just as a jump between energy levels in an atom results in a release or absorption of energy, even minor improvements or actions in the workplace can cascade into larger, transformative impacts on productivity and efficiency. This integration of fundamental science into applied physics was pivotal for the success of the project.

Each individual and component, whether theoretical physics, material procurement, or experimental testing, was a critical element within the larger system that culminated in the first nuclear detonation. This monumental achievement demonstrates how systems harness small, specialized actions to produce an outcome of immense impact. The scientific principles underlying the atomic bomb, nuclear fission, chain reactions, and energy release, parallel the concept of leveraging micro–actions in productivity systems. Each "atomic" decision or adjustment within a system has the potential to trigger a cascade of transformative effects.

Micro-Actions in Macro-Systems

Even the smallest actions have systemic impacts. A seemingly minor adjustment in communication workflows, for example, can ripple through an entire team. Deming's principle of interdependence emphasizes the need for synergy between parts of a system to achieve optimized productivity. This section discusses practical examples where aligning micro-actions with systemic goals led to enhanced operational outcomes.

To draw a scientific parallel, consider a chain reaction within an atomic bomb. The splitting of a single atom unleashes immense energy, triggering a cascade of reactions. Similarly, small, consistent actions within a well-structured system, such as improving response times in customer service, can spark transformative changes across an organization.

Your Atomic Moment

Right now, in your organization, there are dozens of atomic actions waiting to be discovered. They're hiding in plain sight:

- The 30-second interaction that prevents hours of confusion
- The simple tool placement that eliminates daily friction
- The brief question that unlocks innovation
- The small recognition that energizes entire teams

By the end of this book, you'll have:

- A method for identifying these atomic actions (The ATOM Framework)
- A formula for calculating their multiplication potential (CRC)
- A system for implementing without overwhelming your organization
- Real examples from organizations that transformed through atomic thinking

But first, we need to understand why most productivity efforts fail. It's not because people don't work hard enough. It's because we've been blaming the workers instead of fixing the systems.

Your First Atomic Challenge: Before reading Chapter 2, think of one small action in your workday that seems to have effects beyond its apparent importance. Write it down. We'll return to it later.

The Red Bead Experiment and Willing Workers

Understanding variation and the myth of individual performance

One of the most profound lessons I learned during my doctoral studies came from revisiting Deming's Red Bead Experiment. If you're not familiar with it, buckle up. This experiment will fundamentally change how you view performance, productivity, and the role of workers in any system.

The Setup

Picture this: you're running a factory that produces white beads. That's your only job, to produce white beads. But here's the catch: your raw material contains both white beads (good) and red beads (defects). Your workers use a paddle with fifth indentations to scoop beads from a container. The goal? Get only white beads.

Deming ran this experiment with volunteers, assigning roles:

- Line workers (usually six people)
- Inspectors (two people)
- A chief inspector
- A recorder
- The rules were simple but strict:
- Workers must use the approved paddle
- They must follow the exact prescribed motions
- No variations allowed
- Resignations are not permitted (remember this)

What Happens Every Time

Here's where it gets interesting. No matter who plays the workers, no matter how hard they try, the results are predictable:

- Some workers consistently produce fewer red beads.
- Some produce more.
- Management praises the "good" workers.
- Management threatens or fires the "bad" workers.
- Overall system performance never really improves.

Sound familiar? It should. This is exactly what happens in most organizations every single day. Knowing that this is common (keeping in mind the bonuses we give to entice our teams to work harder), and how Deming saw this process as problematic, why do we still do it today?

The Profound Truth

Here is why we still do this today! The ratio of red to white beads is built into the system. No amount of effort, skill, or motivation from the workers can change the fact that red beads exist in the raw material. The workers are "willing workers": they want to do well, and they follow the procedures exactly. Yet their performance varies. We will cover more on willing workers in a later chapter. Why does the performance vary when workers inherently want to do a good job? Well, Deming had a point. The vast majority of performance variation comes from the system, not the individual.

This is the profound truth; however, we often point to the people as the gap. In the words of a past president I worked for, "It's either training or choice" as to why people are not achieving the outcomes expected of them. But what if it is neither? What if something is broken in what we do?

The Willing Worker Concept

The "willing worker" is one of Deming's most important ideas. Deming believed that the vast majority of workers come to work wanting to:

- Do a good job
- Take pride in their work
- Contribute meaningfully
- Go home feeling accomplished

They are willing. They are capable. But too often, they are trapped in systems that make success impossible. Systems that were developed by the people who run the organizations. Systems that are likely limited in performance yet our mental models have us think that what is designed, with intent, is what will produce the highest levels of return. The idea was not that people need external rewards or punishment, but that most employees already want to succeed.

In his words, "A bad system will beat a good person every time." That quote is more than a critique; it's a challenge to leaders. If your workers are willing, then your job is not to "fix" them. It is to unleash them.

The Quantum Perspective: Potential Energy in the Worker

In quantum physics, particles hold potential energy based on their state and position. Likewise, every worker carries within them unrealized potential: ideas, effort, and creativity. The right system can release that energy. The wrong one keeps it inert.

The willing worker is like an electron: full of charge, possibility, and value, but highly dependent on the environment to determine whether that energy is unleashed.

I saw this dramatically during COVID. When systems were simplified - when the focus became just "get groceries on shelves" - willing workers suddenly had space to show their capability. They innovated, collaborated, and exceeded all expectations. The workers didn't change. The system constraints did.

During my dissertation research, I interviewed dozens of grocery workers. You know what I heard over and over? Variations of these:

"I want to help customers, but I don't have the tools."

"I know how to fix this problem, but nobody listens."

"I'm doing the job of three people and still getting blamed for delays."

These weren't lazy workers. These weren't incompetent workers. These were willing workers trapped in broken systems. Technologies were being rolled out with a strong purpose, yet consideration for reality was missing.

The COVID-19 Test Case

Now let's think of a true reality, one in which we all lived. The pandemic provided a natural experiment and case study in system impact. In early 2020, grocery workers were suddenly "heroes." They got bonuses, public praise, and additional support. Some of the unrealistic expectations went out the window, and they "just got to do their jobs." The work didn't change, but the system around them did:

- More staff hired to meet demand, often without scrutiny or pickiness
- Simplified procedures (just get product on shelves)
- Reduced bureaucracy
- Clear priorities
- Public appreciation
- Etc. etc. etc.

What happened? Productivity soared. Morale improved. The same workers who had been "struggling" were now excelling. Performance bonuses went through the roof, and so too did the profits for the organizations these heroes supported. Then, as the pandemic wore on, what changed?

21

- Bonuses ended
- Staff were cut
- Old procedures returned
- Multiple, conflicting priorities emerged
- Public patience wore thin

The workers didn't change. The system did, and performance followed the system, not the workers. Now, fast forward five years, and the "old way" is back. The performance is heavily scrutinized, and the people are back to square one. They are tired, worn out, and underappreciated. Something we can easily reference as one of the major causes of The Great Resignation.

Breaking Down the Performance Myth

Now, Deming was not a fan of quite a few things, including the concept of performance reviews. Let's destroy some common myths about performance:

Myth 1: Good workers succeed regardless of the system.

Reality: Even exceptional workers can only compensate for system failures to a point. Eventually, the system wins.

Myth 2: Performance problems are usually people problems.

Reality: Deming estimated that 94 percent of problems come from the system, and only 6 percent come from individuals.

Myth 3: You can motivate your way to better performance.

Reality: Motivation without system support leads to burnout, not breakthrough.

Myth 4: Competition between workers improves overall performance.

Reality: Internal competition often suboptimizes the whole system.

Building on Deming, these myths are what I believe create some of our problems, barriers, and opportunities today. I once read a book titled *The Nine Lies About Work*, and one of the lies (#6) is the concept of people being able to reliably rate other people. After reading this book, I became fixated on this lie. This framed my thoughts on work. Is what I do today adequately measured? Or, better yet, tying it into this linear thinking, am I being measured on all the things outside of my linear job duties?

The Atomic Connection

So how does all this connect to the concept of the atom or atomic actions? It's rather simple:

If workers are willing and systems determine outcomes, then the highest-leverage actions are those that improve the system, not those that try to "fix" workers. But what if we look at atomic actions?

An atomic action might be any of the following:

- Removing a redundant approval step
- Providing better tools
- Clarifying priorities
- Eliminating conflicting metrics
- Creating information flow

These system improvements multiply the efforts of every willing worker. This is the profound shift in thinking. Instead of trying to split the atom of human performance, breaking down workers to find what's "wrong" with them, we should examine the molecular structure of our systems. Just as atoms combine to create powerful compounds, small system improvements combine to create environments where willing workers naturally excel.

The atomic action isn't about making workers work harder or smarter. It's about removing the friction (more on this later) that prevents their natural capability from flowing. When you fix the system, you don't just solve today's problem, you prevent tomorrows.

This is how small actions create massive impact. Not through heroic individual effort, but through intelligent system design that unleashes the potential already present in every willing worker.

Real Examples of System vs. Worker

Let me share some real examples from my research and lived experiences comparing the system and worker.

The Deli Counter Disaster:

A grocery deli was consistently getting customer complaints. What was management's solution? More training for the workers. Performance reviews. Threats of termination.

What was the real problem? The slicer was old and took twice as long as it should. The scale was in the wrong place, requiring extra steps. The scheduling system puts one person alone during lunch rush. As a store manager at Kroger, I saw this happen. I often received these very same customer complaints, either by phone call or by customer satisfaction surveys. My solution (taught in a rigorous management training program) was to coach and develop. Look at the people, the person working, the deli clerk. But, the real solution could have and should have been through the lens of an atomic action: purchasing a new slicer, relocating scale, or adjusting the schedule. Use the complaints as a point of reference to the system, not the workers. Heck, the workers came to work, meaning they were "willing."

The Checkout Champion

One store has a cashier who is legendarily fast, with metrics such as produce scanning and items per minute at the top of the charts. Management wanted everyone to be like "Sarah." They had her train others. They created competitions. They offered bonuses for speed, etc. But when it came to overall performance, nothing worked. Why? Sarah worked register #6, which had the following:

- The newest scanner
- The best bagging area layout
- The shortest distance to handle coupons
- The most experienced baggers scheduled with her
- Smaller baskets sizes (since she worked the opening shift)

When we think of this, Sarah had the best system to perform. Now, I don't want to take anything away from Sarah. She earned everything she had. She had been with the company for more than twenty years in that same cashier role. If management were to optimize the system for everyone else and put other cashiers on register #6, schedule a strong bagger, and let them open for the day, guess what? They can all become "champions."

25

The Resignation Connection

As part of my dissertation research, I focused on the phenomena of the COVID-19 Pandemic and The Great Resignation. During The Great Resignation, millions of workers didn't suddenly become lazy or entitled, regardless of the subsidies given by the government. They became aware of the following:

- The system was broken
- They were being blamed for system failures
- Other systems (other employers) might be better
- Their willingness was being exploited, not supported

One worker told me, "COVID showed me that they could make the job bearable when they wanted to. When they took it all away, I knew it was a choice. So, I made a choice too: I left." As I mentioned earlier, the pandemic was an eye-opener to the possibilities of a strong system. We reverted back to what our mental models told us "worked." This was likely due to the way the system was designed, right? Since the metrics, dashboards, and tools weren't necessarily built for crisis, when a sense of normalcy arose, shifting back to comfort was easy.

Applying Red Bead Thinking

So, we know the possibilities. We have seen success, high morale, and happiness. But I often wonder why many companies don't apply Deming's findings and frameworks. What happens if we apply his red bead experiment to our organizations today? Here's how to apply these insights:

When performance varies, look first at the system:

    a. What tools are provided?

    b. What processes are required?

    c. What barriers exist?

    d. What conflicting demands are present?

Assume workers are willing until proven otherwise:

    a. Most performance issues aren't motivation issues.

    b. Most workers want to succeed.

    c. Most problems have system roots.

Track system metrics, not just individual metrics:

    a. How long does the process take?

    b. Where do delays occur?

    c. What prevents smooth flow?

    d. What causes rework?

Involve workers in system improvement:

    a. They know where the red beads are.

    b. They've developed workarounds.

    c. They see the waste.

    d. They have solutions.

Asking these questions and applying these insights prepares us to find the atomic actions that will lead us to success.

Your Red Bead Audit

Now, take a moment to think about your own work:

- What "red beads" are built into your system?
- What variation do you see that gets blamed on people but probably comes from the system?
- Where are willing workers being constrained by system failures?
- What one system change could multiply everyone's effectiveness?

Remember: your people aren't the problem. They're willing workers trapped in systems full of red beads. Your job is to reduce the red beads, not blame workers for finding them.

The atomic impact comes from fixing systems, not people. Because when you improve the system, you improve every worker's ability to succeed. That's exponential. That's transformative. That's atomic.

The Profound Shift

We discussed the profound truth. Now lets discuss the profound shift. When you truly internalize the Red Bead Experiment, understanding it deeply, everything changes:

- You stop blaming and start investigating.
- You stop motivating and start enabling.
- You stop managing people and start managing systems.
- You stop accepting variation and start understanding it (more on variation a little later).

This is the foundation of *Atomic Impact* and what I call atomic productivity. Now, Leo Black has a book called *Atomic Productivity* (great book as I mentioned earlier) where he defines atomic productivity as the tiny consistent behaviors or atomic actions that help you reprogram your brain to enjoy tough tasks. This is similar to James Clear's *Atomic Habits*. This is not where *Atomic Impact* takes you. We are not looking at the individual. We aren't looking at the behaviors of people, although we will consider them. We are looking at the organization first. What can we see and affect? When you see systems clearly, you can identify crucial atomic actions that can transform everything. These are the small interventions within organizational systems that create chain reactions of productivity far exceeding their initial investment.

CHAPTER 3

Variation – The Voice of Your System

Rethinking variation as information, not error

Russ Ackoff, a pioneer in systems thinking and organizational theory, often emphasized that one of the most profound missteps in management is mistaking the parts for the whole. He argued that optimizing components in isolation inevitably suboptimizes the system as a whole. Now, Ackoff is known as being somewhat critical of an overly mechanistic approach to systems thinking. In this light, variation, often seen as a nuisance or noise, becomes one of the most important signals a system can produce, therefore, positing that not all variation is bad, a core belief of Ackoff.

The Automobile Analogy

One day I was watching a YouTube video of Ackoff lecturing on systems, and he used an example of an automobile. His thoughts went to this: if I bring a car into this room and take it apart, piece-by-piece, we will have the makings of a car, but not a car itself, essentially no system. This shows the importance of subsystems and the role(s) each part plays. Independently, the wheels of the car, the alternator, the gas tank, etc. etc. cannot make the car go without another component that is connected to that one.

You are probably wondering what this has to do with variation. In the world of management, when something doesn't work or a metric isn't achieved, we look for the "root cause," or the problem, and think we need to fix this. This is usually drilled down to a singular area within the process or system that is blamed for the lack of achieving expectations. However, let's go back to the automobile scenario: we can't blame the wheels for the car not working if they aren't properly connected. Now, don't forget that sometimes, and more often than not, we blame the people!

The Quantum Reality of Variation

Now let's reference back to the atomic consideration of this topic. In atomic theory and quantum mechanics, variation is not a flaw, it is a feature. Electrons do not follow deterministic paths but rather probabilistic distributions. The Heisenberg Uncertainty Principle prompts us that even at the smallest scales, there is always ambiguity in measurement. You can't simultaneously know an electron's exact position and velocity, so we must consider what that means when it comes to expectations of that electron. In essence, variation isn't a limitation of tools, resources, or people; it's a truth of the universe.

So, I leave you with a question on this. Why, in our organizations, do we expect precision where the natural world promises uncertainty?

Rethinking Variation as Information

Since Ackoff saw variation as not all bad, reminding us that variation is not merely something to be reduced, that it is something to be understood, we must rethink variation and retool our minds to see it as information. Deming and Walter Shewhart taught us to distinguish between common cause variation (normal system noise/errors) and special-cause variation (indicators of a shift or anomaly). Shewhart, technically the pioneer behind defining variation, was critical in not pushing what isn't there (regarding variation). In essence, don't force an act on variation unless there is a signal that gives us information to act. This takes me to my world of work. When making "data-driven decisions," we sometimes do this very thing. We force the data to tell us what we want to hear. We go against what Shewhart says not to do.

Let me give you an example. I wanted to find a way to test this theory of atomic productivity. I'll share my definition later. Rather than observing, coming up with a hypothesis, answering questions, etc., I created computational simulations using programs like R, Node.js, etc. My computational models revealed something deeper: variation patterns contain predictive information about system health, improvement potential, and where atomic actions might have the highest impact. When I analyzed simulation data across thousands of organizational scenarios, variation wasn't just noise to filter out; it was a signal to decode, a signal telling me to find the problem.

The False Promise of Uniformity

Uniformity has long been mistaken for excellence. Heck, we even have the field of "Operational Excellence". In many organizations, sameness is equated with control, predictability, and quality. But this mindset reveals a fundamental misunderstanding of how systems behave in the real world. True excellence does not arise from eliminating variation, it emerges from understanding it, designing for it, and performing consistently in spite of it.

Genichi Taguchi, a pioneering Japanese engineer (1924–2012), emphasized that quality lies not in the absence of variation, but in the ability of a system or product to function reliably in the presence of it. Taguchi's work, developed in parallel with that of Deming, challenged the conventional approach of treating variation as a problem to be stamped out. Instead, he argued that robustness, the capacity of a system to perform under diverse and sometimes adverse conditions, is the real indicator of quality.

The Taguchi Loss Function

While the two men did not collaborate directly, their philosophies seemingly complemented each other. Deming focused on the system as a whole, warning against the dangers of reacting to every fluctuation in output (variation), particularly when the variation was systemic (common cause). Taguchi, coming from an engineering perspective, pushed for product and process designs that were inherently less sensitive to noise (variation), meaning they would perform consistently regardless of uncontrollable variables like environmental shifts, manufacturing inconsistencies, or user behavior.

Now, this concept of robustness offers a powerful rebuttal to the false promise of uniformity. You were probably wondering when I would expand here. A system that only functions under ideal, tightly controlled conditions is fragile, not excellent. Think of it this way. If a system is "so perfect" because we assume uniformity helps us isolate our problems, then why do we have the problems in the first place? A robust system thrives in the real world, where variation is inevitable, and change is constant.

Excellence Through Robustness

In short, excellence, then, is not a straight line but a wide path, one that absorbs variation, flexes with uncertainty, and still arrives at its destination. This is echoed in quantum decoherence, the moment when a particle becomes entangled with its environment and collapses into a single state. If our systems are too tightly controlled, they may collapse under real-world complexity. Like quantum systems, we must design for resilience, not perfection. See what I did there? I added some of the physics concepts to show you how this book is different.

Take for example some of the companies we once knew as juggernauts, companies we thought would be here through the end of the world. Blockbuster, a brand known for its tightly controlled model built for a predictable environment, assuming customer behavior and technology would remain stable. When variation emerged (streaming, shifting customer habits, and platform-driven content distribution), the system was too rigid to adapt. In comes Netflix, a company that embraced variation as part of its design. First it offered mail-to-order DVDs, and then it focused on streaming services. Now it owns its own studios to meet customer habits and demands. Obviously, data has helped drive this, but Blockbuster had every chance to meet this same demand. If I remember correctly, they even tried.

Another example of this failure to adapt was Sears, Roebuck & Co. A retail company, a dominant force in catalog and brick-and-mortar sales that once grossed over $53B in revenue in 2006, is no longer with us. Sears failed to adapt its tightly integrated legacy systems and supply chain to meet the demands of this new era we are in today. Rather than redesigning for resilience, Sears tried to preserve a decaying structure, unable to absorb variation from e-commerce, fast fashion, and changing consumer expectations. Its collapse was not from a lack of assets but from a lack of adaptability. This happens every day. We think our systems are untouchable. We assume our processes are excellent. Oh, and to be a high-level market share owner, we assume we have the backing to keep us here long term.

Managing the Spectrum, Not Just the Average

One of the great fallacies in measurement is the average. Ackoff famously warned, "The average American has one breast and one testicle." The point: averages obscure useful data. They tell you very little about the system's real performance or its edge cases. We know that this fun quote is not truth to reality; however, we look at things on an average and even sometimes an aggregate level to tell us what we should do or where we should go.

In workforce management, if 90 percent of employees meet targets, it's tempting to assume the system is working. But what about the 10 percent? They may reveal the following:

- Systemic breakdowns
- Inequities in scheduling
- Unclear policies
- Misaligned tools

Today, I manage a group of targets that aggregate to a total target. Often, we could achieve the 100 percent expectation as a whole but only five to six locations actually carried the entire area. As long as the top was at 100 percent, we were doing what was expected or the system was "working." This aggregate or average total at 100 percent keeps the scrutiny away and the assumptions at bay. Fun way to look at performance, right?

This is related to Price's Law, which states that 50 percent of the work is done by the square root of the total number of people participating. This is called the *domain*. In a team of 100, just ten people do half the work. But here's the critical question: is that because those ten are exceptional or because the system only enables 10 to be productive?

When you tie in Ackoff's fun quote and consider Price's Law, why do we become comfortable with that top line number? Why do we accept the average?

Building Systems that Breathe

Rather than controlling variation, we can design systems that breathe with it. Here's what that looks like:

During my career at Kroger, I was embedded in a system that relied heavily on forecasting. This forecasting enabled the store to adjust its labor/workforce based on forecasted sales, taking into account weather patterns (hurricane season), holidays, school, etc. This is a breathing system. Accepting what is out of our control and using it to optimize expectations.

Rather than rigid procedures, breathing systems use frameworks that allow method variation within outcome parameters. Too often do companies create a plan of best practices and procedures to achieve a goal. Now, consider your job today; the system you operate in. Can you confidently say that your system is breathing or even breathable?

The Quantum Perspective on Organizational Variation

In quantum physics, this would be akin to wavefunction superposition, a state where multiple possibilities coexist simultaneously until context, observation, or interaction with the environment collapses them into a single outcome. Simply put, it's like a coin spinning in the air, representing both heads and tails at once until you catch it and look, and only then does it "decide" which side it is.

If you have ever heard of the experiment of Schrödinger's cat, it is like this. Erwin Schrödinger ran an experiment where he placed a cat in a sealed box with a radioactive atom, a Geiger counter, and a poison vial. If the vial were to break, meaning the atom decayed, the cat would obviously die. If the atom doesn't decay, the cat will survive. Here is the "fun" part of this experiment: the cat is both dead and alive until we open the box and check. This is a superposition of life and death in this scenario. Or, consider Schrödinger's plates. I once saw a Facebook posting where there was a stack of plates in a cabinet with a glass front where the stack of roughly twenty to thirty plates were all leaning against the door. These plates lived in two states, both broken and unbroken. Whether they fell out and broke or simply slid into stability was dependent on when the cabinet door was opened.

Back to the world of the system. Prior to collapse, a system holds potential in many forms (superposition). In this state, all possible outcomes coexist simultaneously, like parallel paths waiting to be chosen. The system has not yet committed to a single trajectory, and its full range of possibilities remains open.

Consider a team facing a production bottleneck. They might implement overtime (traditional path), redesign the workflow (innovative path), or discover an entirely unexpected solution through worker experimentation (emergent path). Until leadership's observation and decision, asking "what's our solution?" and demanding immediate action, all these possibilities exist simultaneously. But here's the critical insight: the very act of measuring and demanding quick resolution often collapses the system into the most familiar, least transformative option.

The Observer Effect in Management

We say that the decision is up to leadership. The leaders make an observation and subsequently a decision. Let's call our leader or manager the observer. The observer doesn't create the outcome from nothing, but their presence and the conditions of observation - what is measured, how it's measured, and in what context - play a defining role in determining which of the many viable paths becomes real. It's not just about seeing; it's about interaction. Observation becomes entanglement, and the act of engaging with the system shapes what emerges.

This principle offers a powerful metaphor for leadership and decision-making. The environments we create, the assumptions we hold, and the metrics we emphasize all influence how potential becomes reality. If we rush to define outcomes too early, we risk collapsing rich possibilities into narrow, rigid results.

This mirrors the decision-making landscape in complex systems. Organizations, like particles in superposition, often contain multiple paths, interpretations, or outcomes within them. The choices leaders make, the measurements they prioritize (KPIs), and the environments they shape all serve to "collapse" those possibilities into one realized state. If systems are overmanaged to force a single path prematurely, based on averages, outdated models, or rigid expectations, they may eliminate valuable possibilities before they are fully understood. My epiphany? Stop forcing things.

A Real-World Example: The NBOT Problem

Let's take how non-billable overtime (NBOT) is measured and often pushed in the realm of contract security. Since a contract is the defining factor for hours of security coverage (physical security guard presence), coverage must be continuous. From a financial perspective, NBOT is the deadliest threat to a security organization's bottom line. This is because most security services offered and filled using overtime that cannot be billed means we are essentially giving away the service.

Often, this number is measured as a percentage of total billed and paid hours and is pushed to be reduced (for obvious financial reasons). When this percentage is looked at, some people, leaders, or managers, push the reduction without intention. They simply reduce it; however, many variables exist that impact this percentage:

- Unexpected absences
- Training requirements
- Travel time between sites
- Client-requested changes
- Security incidents requiring extended coverage

The variation in NBOT tells you about system health. Trying to eliminate it without understanding it is like taking painkillers for a broken bone. You might feel better temporarily, but the underlying problem remains.

The Role of the System Steward

Earlier, I briefly mentioned the assumption that a system works is often driven by the achievement of an average or top-line number. However, it is important to understand that variation is often managed, controlled, or even exiled by those in the field. These are our system stewards. System stewards don't simply lead people; they lead interactions. They recognize that every process, task, and person exist within a network of dependencies. Effective leaders or managers, then, are not heroic problem-solvers, but rather architects of clarity and alignment. They design for flow, not just for control. They work beyond the linearly defined process(es) and work through the variables and feedback loops.

Consider the leader as a gardener rather than a mechanic. A mechanic fixes, replaces, and tightens. A gardener cultivates conditions, tends to systems over time, and understands that some outcomes can't be forced and instead must be nurtured. In a productivity ecosystem, the gardener-leader understands how to create psychological safety, provide direction without micromanaging, and intervene only when the system signals misalignment (variation) when necessary.

What This Means in Practice

Not too long ago, I was playing with Vensim to create a system dynamics workflow for my organization's hiring processes. While most people see this hiring process as linear with a step one (start) through/to step 1,000 (finish), there are variables outside this linear process that create feedback loops and are often where the system stewards manage.

For example, here is the hiring process:

Requisition is created > applicant applies > applicant is interviewed > offer is made > background check is processed > New Employee Orientation (NEO) is scheduled and completed > new hire goes to on–the–job training, etc.

When the NEO session is completed, the new hire has now become part of the workforce and counts toward a designated weekly hiring target. This target tells us our success or failure for the week (KPI). When the target is missed, we often look for the why and directly go to the linear process to see where there was a gap or miss. Was there a variation? Did we not do something we should have done, etc.?

Well, consider the variables outside of the linear process that the system stewards are managing. This might be any of the following:

- System failures in the application portal
- Candidates not showing up to their scheduled NEO sessions
- Background check delays
- Competition from other employers
- Transportation issues
- Weather
- Childcare challenges

These outside variables are often missed or not evenly considered in the context of success for that week. Does this mean we are not aiming to use the variations that exist to determine how to course correct? I think not. It is simply that we are limited by our mental models and where we want to look.

Practical Design Principles for Variation

Now let's look at some practical design principles for variation. Since we know that it is considered noise but should be considered the orchestra's maestro, think through these few principles:

Decentralize sense-making: let those closest to the variation analyze and respond. In my research and experience, the best-performing departments had team leads who could make real-time adjustments without seeking multiple approvals. They understood their variation patterns better than any distant manager could.

Visualize the edge cases: Don't manage to the average; manage to the variance. Create dashboards that show the full distribution, not just the mean.

Build adaptive rules: Instead of fixed policies, use thresholds and triggers. Rather than "Everyone must achieve 95 percent on-time delivery," try "When on-time delivery drops below 95 percent, initiate root cause analysis." The first punishes variation; the second learns from it.

Embed variation-friendly metrics: Instead of only tracking totals, track ranges, stability, and responsiveness.

The Voice of Your System

Ackoff helped us see that a system is not just a collection of parts, it's a set of relationships, similar to those that our system stewards are managing. Variation is how those relationships speak. Whether in human systems or subatomic ones, noise is often just a signal waiting for interpretation.

Your Variation Audit

Take a moment to consider how you look at, find, and perceive variation. Let's call this an audit.

- What variation in your system do you currently try to eliminate?
- What might that variation be trying to tell you?
- Where do you manage to the average instead of understanding the distribution?
- How could you redesign one process to embrace rather than eliminate variation?

In the end, variation is not our enemy. Misunderstanding it is.

When you learn to hear variation as the voice of your system, you can identify those atomic actions that don't just reduce problems but transform possibilities. What looks like noise may be your system screaming the solution.

The Connection Point

Understanding variation as information rather than error was transformative,, but it raised a bigger question: If small variations could tell us so much about our systems, what about small actions? Could they have impacts far beyond their size?

This question consumed me. The more I thought about it, the more I saw connections to the quantum physics concepts I'd been studying. Energy comes in packets. It's not continuous. What if productivity worked the same way?

The next chapter tells the story of how these ideas crystallized into the Atomic Productivity Theory. It wasn't a linear journey. It was messy, frustrating, and full of false starts. When the pieces finally came together, they revealed something profound: we've been thinking about productivity all wrong.

We've been trying to create big impacts with big actions. But what if the secret was the opposite? What if the smallest actions, properly positioned within the right systems, could create chain reactions of transformation?

The answer to that question changed everything.

PART II:

The Atomic Approach

CHAPTER 4

The Birth of Atomic Productivity Theory

From quantum physics to organizational transformation

Now I mentioned that I have a theory: The Atomic Productivity Theory. The Atomic Productivity Theory didn't emerge in a conference room or from a management textbook. It was born from frustration, curiosity, and an unexpected connection between nuclear physics and the very real problem of doing more with less. With AI, technology insertion, process improvements, and operational excellence, we are dealing with a world that wants to and must do more with less. We saw this during The Great Resignation and the Pandemic. But what if we had a solution or even a theory to help us truly achieve our expectations of doing more with less successfully?

The Perfect Storm

In 2023, I found myself at an intersection of experiences that would reshape how I think about productivity.

I had just spent four years researching The Great Resignation and COVID-19's impact on workforce productivity.

My dissertation data showed a disturbing pattern: organizations were demanding more output with fewer resources. Traditional productivity approaches were failing spectacularly, and workers were burning out at unprecedented rates. Yet, some teams were thriving despite having fewer resources than their struggling counterparts.

What made the difference? That question haunted me. I wrote about these two phenomena and saw that the grocery retail industry was in need of more technology, namely self-serve technology (SST). Were customers and workers ready for this? Would organizations have the mental models that say, 'let's replace X for Y (tech)' and see a better return?

The Oppenheimer Moment

As I mentioned earlier, watching *Oppenheimer* wasn't just entertainment for me. With my father-in-law's thirty-five years at Los Alamos, the Manhattan Project was family history. As I watched it for the third time (my family will tell you I can be obsessive), something clicked.

The atomic bomb wasn't created by using more TNT. It was created by understanding that matter itself could be transformed – that a tiny amount of uranium could release energy equivalent to thousands of tons of conventional explosives. What if productivity worked the same way? Could there be something we were missing in our traditional measurement that could be transformed into smaller actions? Could we actually do more with less and it be so profound that we find a strong sense of understanding? I am used to being asked to add more resources, more "explosives" to solve a problem. However, the additional resources are not often the better solution, better yet, even available.

The Half Price Books Revelation

That thought sent me to Half Price Books, where I found that little paperback on quantum theory. As I struggled through Planck's constant and Bohr's atomic model, a framework began forming:

1. **Energy Is Quantized:** Planck showed that energy comes in discrete packets (quanta), not continuous waves. What if productivity also came in packets, specific actions that release discrete amounts of productive energy?
2. **Chain Reactions Exist:** In nuclear fission, one atom splitting triggers others. What if certain actions in organizations could trigger similar cascades?
3. **Critical Mass Matters:** You need enough fissile material for a sustained reaction. What if organizations needed a critical mass of the right actions?
4. **Control is Essential:** Nuclear reactors use control rods to prevent meltdowns. (I'll give you a control rod example later.) What if organizations needed similar mechanisms to prevent burnout from productivity chain reactions?

Understanding the Control Rod Concept

Before we go further, let me explain this crucial concept of control rods that I'll reference throughout this book. In nuclear reactors, control rods are the difference between power generation and catastrophe. They regulate the chain reaction, keeping it in the sweet spot – powerful but not destructive.

Control rods are physical devices made of materials that absorb neutrons. Think of them as the brakes and accelerator of a nuclear reactor. The more control rods you insert, the fewer neutrons are available to continue the chain reaction. Pull them out, and the reaction accelerates. It's a delicate balance: too much control and your reactor goes cold; too little and you risk meltdown. This is the very example of the infamous Chernobyl accident.

Organizations need the same thing. Without control mechanisms, even positive changes can spiral into chaos, burnout, and systemic failure. Just as nuclear engineers carefully modulate control rods to maintain optimal power output, leaders must carefully regulate the pace and intensity of change to maintain sustainable productivity.

This concept becomes critical when implementing atomic actions. You might discover dozens of high-impact improvements but implementing them all at once would be like pulling out all the control rods: the resulting energy release would destroy rather than power your organization. This is the exact thing that took place in 1986 at Chernobyl. Too many control rods were pulled out at once to bring the power back to a higher level. When that power increased at a rate much faster, the control rods were all placed back into the reactor, which created a problem.

From Theory to Reality

Physics metaphors don't pay the bills. I needed to test this in the real world. Fortunately, my dissertation research had given me the perfect laboratory: grocery retail during a crisis - the pandemic.

Why This Isn't Just Another Framework

I know what you're thinking. Another productivity theory? Another leadership book? Another guy trying to tell executives and managers how to run their business? Haven't we had enough? Let's think about enough.

You've probably hear of or tried one or more of these:

- Lean (eliminate waste)
- Six Sigma (reduce variation)
- Agile (respond to change)
- OKRs (align objectives)
- EOS (entrepreneurial operating system)

Each has value. Each has helped organizations improve. So why do we need the Atomic Productivity Theory? Because all existing frameworks focus on optimization, making what exists work better. Atomic Productivity focuses on multiplication, creating chain reactions that transform everything.

Here's the crucial difference:

- **Optimization** asks: "How can we do this 20 percent better?"
- **Multiplication** asks: "What single change could make everything ten times easier?"

The timing couldn't be more critical. Three convergent forces make atomic productivity essential now:

1. **The AI Revolution**: Technology that can amplify human decisions exponentially, if we know which decisions to amplify.
2. **The Post-Pandemic Workplace**: Distributed teams need multiplication effects more than ever.
3. **The Complexity Crisis**: Organizations are drowning in processes that optimization can't fix.

Atomic Productivity Theory doesn't replace other methodologies; it reveals which improvements will multiply through your existing systems.

I went back to my interview transcripts and observation notes, but this time I was looking for something different. Instead of just documenting problems, I was hunting for atomic actions - small interventions that had created disproportionate results.

The Pattern Emerges

As I went back to my dissertation notes (qualitative), I scoured the feedback from participants and catalogued what I today can be seen as atomic actions; patterns that became clear. These patterns took me outside the normal world of trying to place and group my dissertation findings into themes. Rather than looking for consistent words, I looked for the patterns, the atomic actions that the participants did or would have liked to see:

- **Tiny Investment, Huge Return:** The best atomic actions required minimal time/resources but created cascading benefits.

- **System Amplification:** These actions worked because they aligned with system dynamics, not against them.
- **Human-Centered:** Every successful atomic action respected human psychology and capability.
- **Measurable Impact:** The results weren't theoretical; they showed up in real metrics.

Recognizing these patterns was only the beginning. If atomic actions truly create exponential impact through system dynamics, then we should be able to measure and predict their ripple effects. This realization let me to a crucial question. Could we actually calculate the multiplication factor of these small changes? Could we quantify how a single atomic action cascades through an organization, touching process after process, person after person? This wasn't just academic curiosity, it was about giving leaders a tool to identify which small actions could yield the greatest return. What emerged was a way to calculate what I call the Chain Reaction Coefficient.

Calculating the Chain Reaction Coefficient

So here lies the *Atomic Productivity Theory.* I thought I had something big (and still do), but I needed a way to calculate these atomic actions. How do I account for these tiny things that our willing workers are doing? This led me to develop the Chain Reaction Coefficient (CRC). In the world of physics, a chain reaction is a self–sustaining process that triggers multiple subsequent events. These events then trigger even more events, creating an exponential cascade. In the case of the atomic bomb, a neutron strikes the nucleus of uranium–235 which then splits the nucleus, subsequently releasing more neutrons, which then strikes another nucleus of uranium–235, and so on and so on. These small, atomic actions ultimately create the explosion.

Chain Reaction Coefficient Formula (CRC) = (Primary Benefits + Secondary Benefits × 0.7) × Network Reach / (Direct Investment × 1.5)

Let me break this formula down for you. This formula was developed by looking at some of the simulations I had run. Considering the organizations of today, I wanted a formula that would ultimately account for things we look at and consider that are not looked at together, much like Deming, Shewhart, Ackoff, and Forrester's work. Think of this formula like calculating the "bang for your buck" of any and all workplace actions.

- **Primary Benefits**: The obvious, direct savings or improvements you can easily measure

  - Example: A five–minute huddle saves thirty minutes of confusion later.

- **Secondary Benefits × 0.7**: The ripple effects that are real but harder to prove

  - Example: That same huddle builds trust, improves morale, and spreads knowledge.
  - We multiply by 0.7 (70 percent) because we're being conservative. We know these benefits exist but can't measure them perfectly.

- **Network Reach**: How many people/departments feel the impact

  - Just your work = 1.0

- Your whole team = 1.3
- Multiple departments = 1.7–2.5
  The Bottom Half (What You Put In):

- **Direct Investment × 1.5**: The time/money/effort you spend, plus 50 percent extra

  - We add 50 percent because there are always hidden costs: learning time, coordination, and getting people on board.

The ATOM Method

To make atomic actions practical, I've developed the ATOM Method:

Assess: Identify current friction points and energy drains.
Target: Find the 5% of actions with multiplication potential.
Optimize: Calculate CRC and prioritize highest-impact actions.
Multiply: Design systems that spread atomic actions naturally.
This isn't just some theoretical guide. Here's a real calculation:

Example: The Morning Huddle Atomic Action

- **Primary Benefit**: thirty minutes saved in confusion prevention
- **Secondary Benefits**: Team bonding (fifteen-minute value) + error prevention (twenty-minute value) = 35 min. × 0.7 = 24.5 min
- **Network Reach**: Affects a full eight-person team = 1.3
- **Investment**: 5 minutes × 1.5 overhead = 7.5 minutes

CRC = (30 + 24.5) × 1.3 / 7.5 = 9.4

Any action with CRC > 5 has atomic potential. This morning huddle creates nearly ten times return on the time invested.

Another Example: Let's say you are spending time holding a wonderful Microsoft Teams meeting, and your primary goal is to have your team report on certain KPIs or metrics that you need to check performance on.

Scenario: Weekly one–hour Teams meeting with eight people to review results

Direct Investment:

- 8 people × 1 hour = 8 hours
- With hidden costs (prep, tech issues, context switching): 8 × 1.5 = 12 hours

Primary Benefits:

- Information shared? (Could have been an email)
- Decisions made? (Often just "let's take this offline")
- Problems solved? (Usually just identified)
- **Realistic primary benefit: Maybe one to two hours saved in confusion**

Secondary Benefits:

- Some team connection (but often people are multitasking because we know productivity doesn't happen in a vacuum)
- Some alignment (but was everyone listening?)
- **Maybe 1 hour worth × 0.7 = 0.7 hours**

Network Reach:

- **Network Reach**: Affects a full eight-person team = 1.3

CRC = (2 + 0.7) × 1.0 / 12 = 0.23

This meeting is destroying value! You're investing twelve hours to create less than three hours of benefit.

Now let's redesign this meeting atomically:

Replace with a five-minute daily async check-in (each person posts their wins/blocks in Teams):

- Investment: 8 people × 5 min × 5 days = 3.3 hours weekly × 1.5 = 5 hours
- Primary benefits: Same information shared, faster problem identification = 8 hours saved
- Secondary benefits: Less meeting fatigue, more deep work time = 4 hours × 0.7 = 2.8 hours
- Network reach: 1.3 (others can see and learn)

New CRC = (8 + 2.8) × 1.3 / 5 = 2.8

By going from synchronous meeting to async atomic check-ins, you went from destroying value (0.23) to creating it (2.8) – a 12x improvement!

Bottom line: If your CRC is above 5, you've found something powerful. Above 10? That's transformational. Below 1? You're wasting everyone's time.

The CRC Matrix

Now, since you have a slight understanding of the CRC formula and algorithm, here is a matrix that tells and guides your next steps when calculating your own CRC.

CRC Range	Value Level	Meaning	Action
<0	Toxic	Actively destroying value	Kill it immediately
0 – 0.5	Harmful	Waste disguised as work	Eliminate ASAP
0.5 – 1.0	Draining	Costs more than it creates	Automate or eliminate
1.0 – 2.0	Marginal	Barely breaking even	Minimize or redesign
2.0 – 5.0	Worthwhile	Solid ROI	Keep and prioritize
5.0 – 10.0	Multiplying	Significant Chain Reaction	Prioritize and expand
10.0 – 20.0	Atomic	Transformational	Protect fiercely
>20.0	Exponential	Game-changing leverage	Scale across the Org

The Ultimate CRC

In all my research and application, I've found one action with nearly infinite CRC: teaching others to think in terms of chain reactions.

When people start seeing multiplication effects instead of just direct results, everything changes. They naturally gravitate toward high-CRC actions. They eliminate waste without being told. They create innovations you never imagined. That's the ultimate chain reaction: minds that see systems, not just tasks. And that transformation? It's atomic!!

The Resistance and the Breakthrough

Not everyone will be convinced. When I first presented APT to peers and friends the pushback was immediate:

"This is just working smarter, not harder with fancy words."

"You can't reduce productivity to physics equations."

"Our situation is too complex for simple solutions."

But I ask you one question: "What's one small thing your best performers do that creates outsized results?" Today, If I asked a room of executives, knowing I have done this without telling them what I was looking for, they would all realize that they account for atomic actions without the actual measurement using the CRC. Now you have something to quantify the work of top performers.

The Framework Crystallizes

Through iteration, testing, and refinement, the Atomic Productivity Theory emerged with four core key principles:

1. **The Fission Action Principle:** Not all actions are created equal. Some naturally create chain reactions.
2. **The Chain Reaction Coefficient:** Every action's impact can be calculated and optimized.
3. **The Critical Mass Threshold:** Transformation requires 15–20 percent of actions to be atomic.
4. **The Control Rod Requirement:** Uncontrolled chain reactions lead to burnout, not breakthrough.

These principles represent a fundamental shift in how we approach productivity and change. While traditional methods focus on massive initiatives and heroic efforts, atomic productivity reveals that sustainable transformation comes from identifying and amplifying the right small actions.

Why "Atomic" Matters

Some critics might argue that the atomic metaphor is overwrought. I disagree. I disagree because *Atomic Habits* by James Clear has easily been accepted. Here is the stark difference. Clear talks about accumulation. *Atomic Impact* and the Atomic Productivity Theory talk about chain reactions and multiplicative efforts. Here's a little more on why it matters:

- **It Changes How We See Scale:** Traditional thinking says big problems need big solutions. Atomic thinking says the right small action can release enormous energy.
- **It Emphasizes Precision:** You can't split atoms with a sledgehammer. Atomic actions require careful identification and implementation.
- **It Acknowledges Power:** We're not talking about marginal improvements. We're talking about transformation.
- **It Demands Respect:** Nuclear power can create or destroy. Atomic productivity can transform or burn out.

The Personal Test

Before advocating for others to adopt APT, I tested it on myself. Juggling dissertation writing, full-time work, and family, I was the perfect stressed system. I found myself learning to stay up late, spend time with my family, research, write, and even teach as an adjunct professor. Little did I know. I had atomic actions in place that helped me and still help me be successful. These actions helped me answer the question that a lot of people ask. 'How do you juggle all of this with six kids?"

My atomic actions:

- fifteen-minute morning writing (before work): CRC = 8:1
- five-minute end-of-day setup for tomorrow: CRC = 6:1
- Weekly ten-minute win celebration with family at the dinner table (or wherever we sat to eat): CRC = unmeasurable but vital

Result: Dissertation completed while maintaining work performance and family relationships, not through heroic effort, but through atomic design.

The Five Signs of Atomic Potential

Through my research, I've identified five indicators that an action might be atomic. These five indicators will help you identify the atomic actions in your organization.

1. **The Ripple Test**: Does the action create effects beyond its immediate purpose?
2. **The Resistance Paradox**: Is it surprisingly hard to get people to try something so simple?
3. **The Energy Equation**: Do people have more energy after the action than before?
4. **The Spread Signal**: Do people naturally tell others about it?
5. **The Sustainability Secret**: Does it get easier and more powerful with repetition?

If an action exhibits three or more of these signs, you've likely found an atomic opportunity.

Your Atomic Beginning

The Atomic Productivity Theory isn't about adding more to your plate. It's about identifying what's already on your plate that could create chain reactions. How do you use what you have to be successful? And no... this isn't a leadership book. I'll clarify later.

Start here:

- List your daily actions.
- Identify which ones create secondary benefits.
- Calculate rough CRCs.
- Focus on amplifying high-CRC actions.
- Monitor for system stress.

Remember: $E=mc^2$ showed that enormous energy was locked in tiny amounts of matter. APT shows that enormous productivity is locked in tiny, well-chosen actions.

The atoms are already in your organization. You just need to learn how to split them.

CHAPTER 5

Identifying Your Fission Actions

Finding the atoms that matter in your organization

"How do I know which actions are atomic?" That's the question I think many of you will likely ask me with a mix of excitement and skepticism. After all, if atomic actions are so powerful, why aren't they obvious?

The answer lies in how we've been trained to see work. We focus on the big, the visible, the heroic, but atomic actions are often small, subtle, and systemic. They hide in plain sight. In the movie the *Accountant 2*, Ben Affleck, playing Christian Wolff, made a statement while looking for patterns in cases, pictures, and articles taped to a wall. Christian says, "Most brains seek a pattern that's familiar." This is our problem today. We look for familiarity. This familiarity drives comfort, which we all know leads to complacency. Complacency breeds blind spots that competitors exploit. In the workplace, this cycle often manifests as follows:

- **Established processes go unquestioned** – "We've always done it this way" becomes the default response.
- **Skills plateau** – Once-sharp expertise becomes outdated as industries evolve.
- **Innovation stagnates** – Teams stop looking for better solutions because current ones feel adequate.
- **Risk awareness fades** – Routine tasks lose their perceived dangers, leading to safety incidents or security breaches.

The antidote is deliberate disruption of comfort: rotating responsibilities, seeking fresh perspectives from newcomers, scheduling regular "challenge assumptions" sessions, or implementing continuous learning requirements.

The most successful organizations build in systematic discomfort, forcing themselves to question what's working before it stops working. As Andy Grove put it: "Success breeds complacency. Complacency breeds failure. Only the paranoid survive."

The Atomic Action Paradox

Here's the paradox: the most powerful actions in your organization are often the ones that seem insignificant. They're so small that they don't make it into job descriptions, performance reviews, or process documents. Yet they're the difference between systems that sing and systems that struggle.

During my dissertation research, I observed two grocery departments side by side. Same store. Same customer base. Same products. One thrived; one barely survived.

The difference? The thriving department's leader did something that took thirty seconds at the start of each shift. She'd look each team member in the eye and ask, "What's your win for today going to be?"

That's it. Thirty seconds. But it created a chain reaction:

- Employees thought about success, not just tasks.
- They set personal goals.
- They felt seen and valued.
- They helped each other achieve their wins.
- The positive energy infected customer interactions.
- CRC = approximately 12:1

What about the struggling department's leader? He started each shift by listing everything that went wrong yesterday. While yesterday had passed, and the focus was on today, I have seen task lists, etc. cloud the judgement of the actions that could make a noticeable difference for today.

This is my struggle. We are often focused in a reactive state. We often look for what did or did not work and what did or didn't go wrong, and we base our go-forward on those "understandings". While a small atomic action like this thirty–second question might seem cliché, it's within the realm of a transformational, atomic action based on its CRC score.

The Anatomy of a Fission Action

Not every small action is atomic. Fission actions have specific characteristics. In physics, a fission action refers to the splitting of a heavy atomic nucleus. (Uranium–235 is considered heavy.) The chain reaction we spoke of earlier is the result of a fission action. Now here are four considerations of fission actions in the realm of APT:

1. **Natural Multiplication:** They create secondary effects without additional effort. The morning huddle doesn't just share information; it builds trust, prevents errors, and energizes teams.
2. **System Alignment:** They work with natural system dynamics, not against them. A fission action in one organization might fizzle in another if the systems are different.
3. **Low Activation Energy:** Like catalysts in chemistry, they require minimal energy to start but release significant energy once triggered.
4. **Sustainable Reaction:** They can be repeated without degrading. In fact, many fission actions become more powerful with repetition. Think of this as building behaviors.

The Five Types of Fission Actions

Through my research and computational simulations, I've identified five categories of fission actions:

Type 1: Communication Catalysts

These actions dramatically improve information flow.

Example: The "Two-Minute Download"

A software team implemented a practice where anyone learning something new had two minutes at the next standup to teach it.

Investment: two minutes

Result: 70 percent reduction in repeated mistakes, 40 percent faster onboarding for new team members

CRC: 8:1

Type 2: Decision Accelerators

These actions speed up decision-making without sacrificing quality.

Example: The "$100 Rule"

A retail manager gave every employee authority to solve any customer problem costing less than $100 without approval.

Investment: One announcement

Result: 90 percent faster complaint resolution, 60 percent increase in customer satisfaction

CRC: 15:1

Type 3: Energy Multipliers

These actions boost human energy and motivation.

Example: The "Victory Lap"

A warehouse team started each shift by having yesterday's top performer share their method.

Investment: 3 minutes

Result: 25 percent productivity increase, 50 percent reduction in safety incidents

CRC: 10:1

Type 4: Friction Eliminators

These actions remove small barriers that create big delays.

Example: The "Supply Station"

A nurse manager created supply caches at three points in the ward instead of one central location.

Investment: one hour setup, five minutes daily restock

Result: Saved forty-five minutes per shift in walking time, 30 percent more patient interaction

CRC: 9:1

Type 5: Learning Loops

These actions accelerate skill development and knowledge transfer.

Example: The "Teach-Back"

After any training, employees had to teach the concept to someone else within twenty-four hours.

Investment: ten minutes

Result: 85 percent retention vs. 20 percent for traditional training

CRC: 20:1

These five types of fission actions create tremendous value, but here's what's remarkable: the most powerful atomic actions in your organization might be so small they're nearly invisible. They don't fit neatly into categories because they're subtle behaviors that seem insignificant until you calculate their impact.

Now, consider these nearly invisible fission actions that transform everything they touch:

The Pause

A customer service manager trained her team to pause for two seconds before responding to angry customers.

Investment: 2 seconds

Result: 40 percent reduction in escalations

Why it works: Breaks the anger-defensiveness cycle CRC; Unmeasurable but transformative

The Round

An ICU doctor started making rounds in the same order every day.

Investment: 0 (just consistency)

Result: 25 percent reduction in missed medications

Why it works: Creates predictable patterns nurses could plan around CRC; Infinite (no cost, massive benefit)

The Question

A project manager ended every meeting with "What's one thing we didn't discuss that we should have?"

Investment: thirty seconds

Result: Caught 60% of potential problems before they manifested Why it works: Creates psychological safety for raising concerns CRC: 50:1 or higher

These examples prove that atomic actions don't need to be complex or costly. Sometimes the smallest adjustment, a pause, a pattern, a question, creates the biggest chain reaction.

The Fission Action Audit

So how do you find the fission actions hiding in your organization? After all, these actions already live within your organization. They are simply unaccounted for.

Step 1: Shadow your "stars," and spend a day with your highest performers. Don't watch their official activities; watch the tiny things.

Step 2: Map the multipliers for every action you observe, and ask:

- What else does this trigger?
- Who else benefits?
- What problems does it prevent?
- What opportunities does it create?

Step 3: Calculate the energy using the CRC formula, but don't get hung up on precision. Even rough calculations reveal which actions pack atomic punch.

Step 4: Test for transferability. Can others replicate this action? The best fission actions are simple enough to transfer but powerful enough to matter.

The Multiplication Method

Once you've identified fission actions, how do you multiply their impact? Remember: APT isn't about doing more things (accumulation). It's about doing things that trigger chain reactions (multiplication). The Chain Reaction Coefficient captures this multiplicative effect, showing which actions create exponential rather than linear impact.

1. **Start with Natural Adopters**: Find people already inclined toward the action. Success stories accelerate adoption.
2. **Make It Observable**: Fission actions spread when people can see their impact. Create visibility.
3. **Lower the Barrier:** Remove any friction to attempt the action. The easier to try, the faster it spreads.
4. **Celebrate Early Wins**: Recognition accelerates adoption. Celebrate both the action and its impact.
5. **Build It Into Systems**: Eventually, fission actions should become "how we do things here."

Common Mistakes in Identifying Fission Actions

Now that you have a few ways to multiply your fission actions (remember control rods are necessary), there are mistakes that should be considered and avoided.

Mistake 1: Confusing Big with Powerful

The weekly all-hands meeting might seem important, but does it create chain reactions? Often, the five-minute daily check-in has higher CRC.

Mistake 2: Forcing Unnatural Actions

If an action requires constant reminders or enforcement, it's not truly atomic. Fission actions should feel natural once discovered.

Mistake 3: Ignoring Context

An action that's atomic in one context may be inert in another. Always consider system dynamics.

Mistake 4: Over–Complicating

The best fission actions are stupidly simple. If you need a manual, it's not atomic.

The Compounding Effect

Remember: fission actions compound. One atomic action might have a CRC of 5:1. When you stack multiple fission actions, they don't just add; they multiply. While not the same as compound interest that grows predictably over time, this is more like creating a reactor where each atomic action makes the others more powerful.

Thinking about what I did as a Kroger store manager, I have identified five fission actions:

- Morning energy check/quick lap (CRC 4:1)
- Two-minute teach-backs/coaching (CRC 6:1)
- End-of-day wins sharing/closing huddle (CRC 5:1)
- Rotation reminder system (CRC 3:1)
- "Three before me" rule (CRC 7:1)

Individually, each was powerful. Together, they transformed a struggling operation and gave me the ability to drive stronger performance. I simply did not notice them at the time. The compound CRC? Over 25:1.

Why such dramatic multiplication? Because each action reinforced the others. The morning energy check made people more receptive to peer teaching or answering questions I may not have been available for. Better communication from teach–backs and coaching made the "three before me" rule more effective. Problem-solving authority combined with wins sharing created an innovation spiral.

They didn't just stack benefits; they created an environment where productivity isn't pushed but pulled by the system itself. Each atomic action made the next one more likely to succeed, more likely to spread, and more likely to stick.

That's the power of identifying and implementing multiple fission actions. They create a critical mass where transformation becomes inevitable.

The atoms are already in your organization. Now you know how to find them.

CHAPTER 6

Designing for Efficiency and Innovation

Creating systems that multiply human potential

Design thinking has become a buzzword, thrown around conference rooms like confetti, but most organizations fundamentally misunderstand what design really means in the context of productivity systems.

Design isn't about making things pretty. It's about making systems that work with human nature, not against it. It's about creating environments where atomic actions naturally emerge and multiply.

The Grocery Store Revelation

During my dissertation research, I spent countless hours observing grocery stores. Two stores, with the same company and in the same neighborhood, revealed radically different performance. What was the difference? Design.

Store A was designed for efficiency:

- Logical product flow
- Clear sight lines
- Tools where workers needed them
- Natural customer paths
- Intuitive signage

Store B was designed by someone who'd never worked retail:

- Confusing layout
- Blind spots everywhere
- Tools in central locations (far from use)
- Customer flow that created bottlenecks
- Signage that required interpretation

Same workers. Same customers. Same products. Completely different outcomes.

Store A's design had a system-wide CRC of approximately 3:1. Every action was easier, faster, and more natural. Store B's design had a CRC of 0.3:1. Every action required extra effort, time, and thought. What you are probably wondering is, why is grocery retail so relevant, right? Well let's just say it is the complexity of a system that is not often seen as a system. Between store A and B, there were stark differences. Differences that we don't often acknowledge. These same differences and the psychological drivers that differentiate between the successful and unsuccessful stores can easily translate to any industry today.

The Role of Design Thinking

Design thinking involves empathizing with stakeholders, defining problems, ideating solutions, prototyping, and testing. Here's what most miss: design thinking for productivity isn't about designing solutions. It's about designing systems that generate solutions, systems that enable the users (workers/employees) to effectively use the system to be effective, productive, and efficient.

Let me illustrate with an example. Let's think of a hospital. This hospital was struggling with medication errors. From experience, I have witnessed this take place. When my daughter was born, she was expected to receive a certain vaccination (Hep–B) but was accidentally given a flu shot. Something not intended for infants born only hours earlier. How could this have happened? How do we correct this? The traditional approach is more training, stricter procedures, and punishment for mistakes (Her nurse was sent home).

The design thinking approach is to spend a week with the nurse(s) that made mistakes and evaluate the location of items. After speaking with the leadership of the hospital, we learned that the medications were next to each other, and the vials looked similar. Was this the fault of the nurse? No. It was the system. Taking this into account and thinking through this as I write this book, there must have been a solution to mitigate the error for other families.

What might I suggest as the design solution?

- Meds arranged by frequency of use
- "Do not interrupt" zones with visual indicators
- Focused LED lighting in med rooms

- Large-print labels
- 5S (Sort, Straighten, Shine, Standardize, Sustain)

While these solutions might seem small, they may have a lasting impact on the effectiveness of nurses. While I do not know if these types of solutions were already in place, the potential for accidents can easily be reduced or mitigated.

Design Principles for Atomic Systems

Creating systems that naturally generate atomic actions isn't accidental, it requires intentional design based on how humans actually work, not how we wish they would. Through my research, I've discovered that the most powerful systems share five fundamental design principles. These principles act as a blueprint for leaders who want to stop fighting human nature and start harnessing it, turning everyday operations into engines of exponential improvement

Principle 1: Friction Elimination

Every point of friction in a system reduces CRC. Design should ruthlessly eliminate friction.

Bad design: Employees must log into three systems to complete one task.

Atomic design: Single sign-on with automated task flow.

Principle 2: Natural Action Paths

Humans follow paths of least resistance. Design systems where the easiest path is also the most productive.

Bad design: Safety equipment stored in a locked cabinet across the facility.

Atomic design: Safety equipment at point of use, impossible to start work without it.

Principle 3: Visible Impact
People need to see the results of their actions to stay motivated. Design feedback loops into everything.

Bad design: Monthly reports on performance.

Atomic design: Real-time dashboards showing the impact of current actions.

Principle 4: Error Prevention vs. Error Correction
It's better to make errors impossible than to catch them after.

Bad design: Review process to catch data entry errors.
Atomic design: System that only accepts valid data formats.

Principle 5: Human Energy Optimization
Design should work with human psychology, not against it.

Bad design: Important decisions required at 4 PM when decision fatigue peaks.

Atomic design: Critical thinking tasks scheduled for morning peak performance.

When you apply these five principles together, something remarkable happens: your organization stops depending on heroic individual efforts and starts generating success systematically. This is how atomic actions scale. Not through mandate or motivation, but through intelligent design that makes high performance the natural outcome.

The Atomic Design Process

You may be wondering about the atomic design process. Transforming systems to generate atomic actions requires a systematic approach that balances vision with reality. This five-phase process has consistently produced systems where small improvements create exponential impact, turning organizational friction into fuel for growth.

Phase 1: Current State Mapping

Before designing anything new, understand what exists:

- Shadow users through entire processes.
- Map every touchpoint.
- Identify every friction point.
- Calculate current CRCs.
- Note workarounds, which reveal design failures.

Phase 2: Ideal State Visioning

Without constraints, what would perfect look like?

- Zero friction flow
- Natural action sequences
- Built-in feedback loops

- Energy generating rather than draining
- Self-correcting systems

Phase 3: Constraint Integration
Now add back reality:

- Budget limitations
- Technical constraints
- Regulatory requirements
- Cultural factors
- Change capacity

Phase 4: Atomic Opportunity Identification
Where can small design changes create chain reactions?

- Which friction points affect the most people?
- Which improvements would multiply?
- What's the lowest-hanging fruit with the highest CRC?

Phase 5: Prototype and Test
Start small, and measure impact:

- Pilot with willing early adopters.
- Measure actual vs. expected CRC.
- Iterate based on feedback.
- Scale what works.

These five phases are not and should not be new for many. Much like atomic actions, they are often forgotten, unseen, or not considered. By focusing on these phases (especially Phase 3 "reality"), we can effectively understand where we are and where we need to go, not just where we want to go.

Operational Frameworks Integration

The best designs incorporate proven operational frameworks, but with an atomic twist. These designs take the traditional operational frameworks and add atomic considerations; these operational frameworks aren't still relevant for no reason.

Lean + Atomic Design

Traditional Lean focuses on eliminating waste. Atomic Lean focuses on creating chain reactions that eliminate waste automatically.

Example: Instead of training people to identify waste, design systems where waste is impossible. A parts bin that only holds exactly what's needed for one unit makes overproduction impossible.

Six Sigma + Atomic Design

Traditional Six Sigma reduces variation through control. Atomic Six Sigma designs systems that thrive on beneficial variation while naturally dampening harmful variation. Remember that variation is not always bad!

Example: Instead of rigid scripts for customer service, create frameworks that guide conversations while allowing personality to shine through.

Agile + Atomic Design

Traditional Agile responds to change. Atomic Agile creates systems that generate beneficial change automatically.

Example: Instead of sprint retrospectives that require facilitation, build reflection triggers into daily work that surface improvements naturally.

Technology Integration in Atomic Design

Technology should amplify human capability, not replace it. This was essentially the core of my dissertation. When researching, I knew we were operating in a world that was likely not going back to the pre-COVID era. I focused on the gaps in productivity and asked, "Could we replace humans with self-serve technologies?" After interviewing several people, it was clear that human interaction and human appreciation within a business and within a system are still necessary, at least through the eyes of the customer. Look at it this way; most technological implementation fails because it's bolted onto broken systems. I have witnessed technology be inserted into a process or 'system' only to see it not effectively accepted. Take for example, Fred Davis's *Technology Acceptance Model*. This model is built on the usage of technology, framed through ease of use and usefulness. When technology is integrated, the users will only accept the technology if the usefulness and ease of use is apparent. What happens when we assume technology is the savior, or the answer to our problems (often seen by those not in the field)? It fails.

Successful implementation should be designed to amplify human capability. Take for example, ChatGPT. If used properly, this technology (and all generative AI) should amplify us (humans), not replace us.

The Balance Point

Here's a critical insight: the best system design balances automation with human judgment. Wholeheartedly, I believe that human judgement is the differentiator between humans, and what many people think will become sentient, robots (or AI). There must be a balance, but does a balance require 50/50? Not quite. Take a spoon and balance it on your finger. It will likely not balance at the exact mid-point of the length of the spoon. So, consider this. Pure automation creates brittle systems. Pure human control doesn't scale. The magic is in the mix. If we are going to use technology, we must do so methodically and with intent. This is where the mix comes in. I call this mix the 80/20 Design Rule:

- 80 percent of routine work should be automated or systematized
- 20 percent should require human judgment, creativity, and adaptation

Now there are many references to this 80/20 mix, the most prominent being the Pareto Principle where 80 percent of consequences stem from 20 percent of the causes. Much like Price's Law, there is a domain that leads to a disproportionate level of outputs. However, in the case of this design ratio, this level of balance allows for a few things, things we will call balance, which include the following:

- Prevents job displacement fears
- Maintains system flexibility
- Allows for innovation
- Keeps humans engaged
- Maximizes both efficiency and effectiveness

Real-World Design Transformations

Let's consider how atomic design principles could transform various workplace environments:

The Warehouse Scenario

Imagine a distribution center struggling with fulfillment accuracy. Traditional design might rely on paper pick lists and vendor-based organization. What if we applied atomic thinking?

Consider the possibilities:

- Voice-directed technology could free workers' hands and eyes.
- Organizing products by velocity rather than vendor could reduce travel.
- Zone-based workflows might enable parallel processing.
- Weight verification systems could prevent errors before they cascade.

Each small design change could multiply efficiency while reducing physical strain on workers, workers we know are already stressed or burned out.

The Call Center Environment

Think about a typical customer service center with rigid structures. Traditional cubicles, scripted interactions, and system=controlled breaks often create friction, which we don't like.

An atomic redesign might explore the following:

- Flexible team spaces that adapt to different work styles
- Conversation frameworks that guide without constraining
- Metrics that value problem resolution over speed

- Autonomy in break scheduling within operational needs

These design shifts could theoretically transform both employee experience and customer outcomes.

The Restaurant Kitchen Challenge

Consider a busy kitchen struggling with order flow. Traditional station arrangements based on culinary tradition might not optimize for modern service demands.

Atomic design thinking suggests the following:

- Workflow-based station positioning
- Digital systems that create visual clarity
- Communication methods beyond verbal chaos
- Dynamic menu adaptation to operational capacity

Such changes could theoretically reduce stress while improving service speed and quality.

What is the key insight? Small, thoughtful design changes in any environment can create multiplication effects far beyond their initial investment.

The Meta-Design Principle

Here's the ultimate design principle: design systems that design themselves. This isn't about artificial intelligence or simply applying automation. I mean let's face it. Automation is typically just repetitive tasks being moved from humans that triggers a notification.

The best atomic systems evolve. They learn. They adapt. How?

- Build in feedback mechanisms.

- Create space for experimentation.
- Reward system improvements.
- Make adaptation part of the process.

Example: Let's go back to grocery retail. A grocery store's night crew starts with a simple change: instead of stocking by delivery order, they track which aisles create the most customer complaints about empty shelves. Each stocker carries a pocket notebook, marking just two things: what ran out and when they noticed it. No fancy technology, just pencil and paper.

Within a month, patterns become obvious. Milk always empties by 7 PM on Thursdays. Bananas disappear on Sunday mornings. Friday night means depleted beer shelves. The crew naturally starts adjusting their stocking order to match these patterns, hitting high-depletion items first. Nobody told them to do this; the simple act of tracking revealed the need.

The system evolves further. One stocker notices that cornbread mix sells out whenever collard greens go on sale. He starts pre-staging cornbread near the produce prep area. Another stocker realizes that restocking cereal at 2 AM means it sits untouched until morning, but stocking at 5 AM catches early shoppers. She shifts her routine. These discoveries get shared during shift changes through quick notes on a whiteboard. Soon the entire night operation reorganizes itself around actual depletion patterns rather than corporate planograms. The revolution happens through observation, experimentation, and simple communication. The system teaches itself how to better serve customers.

This example, believe it or not, is reality in the grocery retail industry. While big players (Kroger, Walmart, etc.) have technological solutions to challenges such as these, simple solutions like these are not apparent everywhere.

Your Design Challenge

Let's take this entire chapter into perspective. This week, pick one system you interact with daily. Apply atomic design thinking:

- Map the current state (with all its friction).
- Envision the ideal state (zero friction).
- Identify the highest CRC improvement you could make.
- Prototype it (even if just for yourself).
- Measure the impact.

Remember: great design isn't about perfection. It's about creating systems where imperfect humans can achieve excellent results. That's atomic design, systems that multiply human potential rather than constraining it.

Design determines destiny. Make yours atomic.

PART III

Engineering the Chain Reaction

CHAPTER 7

Building Your Control Rods

*Preventing meltdown while
maximizing chain reactions*

Back in Chapter 4, when I introduced the Atomic Productivity Theory, I mentioned that control rods were essential, that without them, even positive chain reactions could spiral into organizational meltdown. Now it's time to dive deep into exactly how to build these critical control mechanisms into your organization.

In nuclear reactors, control rods are the difference between power generation and catastrophe. They regulate the chain reaction, keeping it in the sweet spot, where it is powerful but not destructive. The same principle applies to organizational productivity. While the concept is simple, the implementation requires careful design and constant attention.

This chapter will show you exactly how to build, calibrate, and maintain the control rods that will keep your atomic productivity sustainable rather than self-destructive.

The Cautionary Tale

Imagine a tech startup discovering atomic productivity principles. An energetic CEO becomes instantly converted to the methodology. Within two weeks, the leadership team identifies and implements 47 high-CRC actions simultaneously.

- Without control mechanisms, here's what would likely unfold:
- Employees overwhelmed by constant change begin resigning
- Remaining staff freeze, unable to prioritize competing improvements
- Customer service collapses as teams struggle with new processes
- Productivity plummets below pre-transformation levels

This scenario illustrates why even beneficial changes need regulation. Implementing dozens of atomic actions without control rods doesn't create transformation, it triggers organizational meltdown. Each individual improvement might have a CRC of 10:1, but combined without restraint, they create negative value.

The principle is simple: atomic reactions require control rods. In organizations, this means pacing change, monitoring stress levels, and building in recovery time. Without these controls, your transformation becomes destruction, not through resistance or failure, but through uncontrolled success.

Understanding Organizational Fission Rate

Every atomic action releases organizational energy. The question is whether that energy powers transformation or triggers explosion. Like uranium in a reactor, the rate of change determines whether you generate sustainable power or create a meltdown. Get this wrong, and even your best improvements become weapons of organizational destruction. That energy can do the following:

- Transform power (controlled reaction)
- Fizzle out (subcritical reaction)
- Explode destructively (uncontrolled reaction)
- The key is maintaining the right fission rate, enough atomic actions to create sustained improvement, but not so many that the system overheats.

Through my research, I've found that optimal rate varies by organization, but follows predictable patterns:

- Stable organizations: 1–2 atomic actions per month
- Growth–mode organizations: 3–4 per month
- Crisis organizations: 5–6 per month (but only temporarily)
- Startup/transformation: Up to 10 per month (with extreme care)

These aren't arbitrary limits, they represent the maximum sustainable rate of change before human systems begin to break down. Exceed them, and you'll see the warning signs: rising error rates, employee burnout, customer complaints, and eventually, complete paralysis. The solution isn't to avoid change but to build mechanisms that regulate it. This is where control rods become essential. Not as brakes on progress, but as the very tools that enable sustainable transformation.

Types of Control Rods

Just as nuclear reactors use different types of control rods for different purposes, some for fine-tuning, others for emergency shutdown, organizations need multiple control mechanisms working in concert. Each type serves a specific function in maintaining the delicate balance between transformation and stability.

Pace Governors

These control the rate of change:

- Implementation calendars (no more than X changes per period)
- Pilot requirements (testing before full rollout)
- Staged deployments (department by department)
- Mandatory stabilization periods between changes

Example: A retail chain implemented a rule: new atomic actions could only launch on the 1st and 15th of each month. This created predictable change windows and prevented initiative fatigue.

Stress Monitors

These detect when the system is overheating:

- Employee pulse surveys
- Burnout indicators (sick days, errors, or conflicts)
- Customer feedback trends
- Performance volatility

Example: A hospital created a "change thermometer," a simple daily poll asking staff to rate their change stress from 1–10. When averages hit 7, all new initiatives paused until it dropped below 5.

Safety Valves

These provide emergency pressure release:

- "Stop the line" authority for any employee
- Temporary reversal options
- Extra support during high-change periods
- Clear escalation paths

Example: A warehouse gave every employee a red card. Playing it immediately paused any new initiative for forty-eight hours while concerns were addressed. It was used three times in the first year and prevented three potential disasters.

Absorption Mechanisms

These help the system process change energy:

- Reflection time built into schedules
- Discussion forums for processing changes
- Additional coaching during transitions
- Buffer capacity for learning curves

Example: A software company implemented "Digestion Fridays," two hours weekly dedicated to practicing new atomic actions, discussing challenges, and sharing adaptations.

The Control Rod Design Process

Just as nuclear engineers must carefully design control mechanisms before starting a reactor, organizations need thoughtfully designed control rods before implementing atomic productivity. The following types and steps will help you create control mechanisms that regulate your transformation without stifling it.

Step 1: Assess Current Capacity

Before implementing any atomic actions, understand your organization's change capacity:

- Recent change history (Are people already saturated?)
- Cultural readiness (Is change seen as positive or threatening?)
- Resource availability (Do you have a buffer for learning curves?)
- System stability (Are basics working well?)

Step 2: Design Detection Mechanisms

You can't control what you can't see. Build in early warning systems:

- Leading indicators (early signs of stress)
- Multiple perspectives (not just manager views)
- Real-time data (not quarterly surveys)
- Qualitative and quantitative measures

Step 3: Create Response Protocols

When warning signs appear, what happens? Pre-plan responses:

- Yellow flags: Slow implementation pace
- Orange flags: Pause new initiatives
- Red flags: Active intervention and support
- Emergency: Full stop and reset

Step 4: Build in Recovery Time

Change is exhausting. Design in restoration:

- Celebration periods after successful implementations
- Stability phases between major changes

- Extra support during high–change periods
- Recognition for adaptation efforts
- Real–World Control Rod Implementation

The Chernobyl Lesson: Why Control Rod Design Matters

On April 26, 1986, the infamous Chernobyl nuclear disaster taught the world a horrifying lesson about control rod design. Much like the design of control rods in your organization must have intent, these control rods at Chernobyl satisfied their intent; however, one fatal flaw existed. The reactor's control rods had graphite tips. When operators scrambled to insert all control rods to slow the fission taking place in the reactor, those graphite tips actually accelerated it. The very mechanisms meant to prevent disaster triggered it instead.

This is precisely what happens in organizations with poorly designed control mechanisms. A stressed organization implements "emergency brakes," such as hiring freezes, innovation moratoriums, and top-down edicts that actually accelerate dysfunction. These control mechanisms themselves become the problem.

I've seen organizational Chernobyls: companies that responded to change overload by centralizing all decisions (graphite tip: destroying local adaptation), freezing all new initiatives (graphite tip: killing momentum and morale), or implementing draconian metrics (graphite tip: driving gaming behaviors). Their control rods didn't regulate the chain reaction; they detonated it.

The lesson is clear: control rods must be designed to actually slow reactions, not accidentally accelerate them. They must be tested under stress, not just normal operations. Most importantly, they must work with your system's nature, not against it.

Your organizational control rods, whether pace governors, stress monitors, safety valves, or absorption mechanisms, must be carefully designed, continuously monitored, and regularly calibrated. When the chain reaction of change accelerates beyond control, you need absolute confidence that your control mechanisms will regulate, not detonate.

Design your control rods thoughtfully. Test them thoroughly. Adjust them regularly. Your organization's sustainable success depends on it.

Building Your Control Rod System

Here's a practical guide to implementing control rods in your organization. Think of this as your pre-flight checklist. Each step builds on the previous one to create a comprehensive safety system for your atomic transformation.

Baseline Assessment Survey team on current change stress (1–10 scale). List all changes from the last six months. Identify signs of change fatigue. Assess available support resources.

Design Controls Set maximum change rate. Create stress indicators. Design response protocols. Build in recovery mechanisms. Keep it simple and visible.

Communication Explain control rods as safety, not limitation. Share the monitoring approach. Clarify escalation paths. Celebrate the commitment to sustainable change.

Implementation Start monitoring systems. Test response protocols with small issues. Adjust based on early feedback. Prepare for the first real test.

Common Control Rod Mistakes

Setting and Forgetting – Control rods need constant adjustment. What works in January might be too restrictive in June.

Manager-Only Monitoring – Managers are often the last to know when teams are overwhelmed. Include peer and self-reporting.

Punishment Focus – If using a control rod is seen as failure, they won't be used until it's too late.

Inflexibility – Sometimes you need to temporarily exceed limits (crisis, opportunity, etc.). Build in override mechanisms with appropriate safeguards.

Once your control rods are functioning, you'll discover something remarkable: the system itself will tell you how it needs to be adjusted. Teams will start suggesting refinements. Patterns will emerge showing when to tighten or loosen controls. This natural evolution points to the next level of control rod sophistication - systems that calibrate themselves.

The Meta-Control System

What is the ultimate control rod? A system that adjusts its own control rods. This requires the following:

- Regular review of control effectiveness
- Feedback loops on the feedback loops
- Authority to adjust limits based on evidence
- Cultural support for sustainable pace

Think of this as organizational self-awareness. The system monitors its own monitoring. Just as your body automatically adjusts breathing rate based on activity, a meta-control system automatically calibrates change capacity based on real-time conditions. When stress indicators trend up, controls tighten. When absorption improves, controls relax. This isn't automated rigidity; it's intelligent adaptation. The system learns what level of change it can sustain and adjusts accordingly, preventing both stagnation and burnout. It's the difference between a thermostat you must manually adjust and one that learns your patterns and preferences - true organizational intelligence.

Your Control Rod Challenge

This week, design one control rod for your atomic productivity efforts:

- Choose a type (pace, stress, safety, or absorption).
- Define clear triggers.
- Design response protocols.
- Test with a small scenario.
- Refine based on results.

Remember: the goal isn't to go slow. It's to go sustainably fast. Like a nuclear reactor, organizational transformation is powerful but needs respect. Control rods ensure you harness the power without the meltdown.

In the end, sustainable transformation beats spectacular burnout every time.

CHAPTER 8

Guiding Small
Acts Through Systems

How to orchestrate atomic actions
for exponential impact

Small actions within an organizational system have exponential potential when systematically guided. As Deming famously noted, "A bad system will beat a good person every time." But flip that around: a great system will multiply every good action exponentially!

The key isn't just identifying atomic actions. It's creating systems that guide, amplify, and sustain them.

The Orchestra Metaphor

Lately, and largely throughout my dissertation, I found myself listening to a new genre of music, classical orchestra. It helped me think and focus, and now, I think I am on a path toward looking for some vinyl records (if my wife can tolerate it).

Think of your organization as an orchestra. Individual musicians can play beautiful notes. These are your atomic actions. Each note matters on its own, but without a conductor and sheet music, you don't get a symphony. You get noise, albeit, noise, with great potential. The same thing happens with atomic actions in your organization.

Without a system to guide these actions, they're all over the place. Some happen, and some don't. Some people do them, others don't bother. It's not that people are resistant; they just don't have structure. You've got all this potential just sitting there, unused.

Here's where it gets interesting. Add what I call a "guidance system," and everything changes. This isn't about controlling people. It's about creating conditions where atomic actions naturally happen and align. Think visual cues at the right moments. Reminders that actually work. People checking in with each other. Easy ways to track progress. Regular celebrations of what's working.

When you put these pieces together, those same scattered actions start working in sync. People go from reluctant to engaged and from random to reliable. The compound ripple coefficient of each action grows because now you have consistency, predictability, and a system that sustains itself.

Feedback Loops: The Heartbeat of Systems

Feedback loops ensure that systems self-correct and adapt to changing conditions. Here's what most organizations get wrong: they create feedback loops for problems, not progress.

The Traditional feedback loop:

Error → Detection → Correction → Monitoring

This is reactive. You wait for something to break and then scramble to fix it. By the time you detect the error, damage is done. Correction takes time and resources. Monitoring feels like surveillance. People hide mistakes rather than learn from them.

The Atomic feedback loop:

Action → Impact visibility → Amplification → Evolution

This is proactive. You start with positive actions and make their impact immediately visible. Success gets amplified through sharing and recognition. The system evolves as people naturally adopt what works. Growth happens through attraction, not correction.

Let me give you an example of an atomic feedback loop. A typical auto manufacturer discovers a door panel alignment issue. Management launches an investigation. They identify the worker responsible, require retraining, and add three new inspection points. Quality control increases random checks. Workers now spend 20 percent more time double-checking their work. The defect tracking system sends daily reports highlighting every flaw. Supervisors patrol the floor looking for mistakes, micromanaging, and looking for who to discipline.

Result: Workers become risk-averse. They slow production to avoid being blamed. When they spot potential improvements, they stay quiet with the thought of "why risk trying something new?" The alignment issue improves marginally, but overall quality stagnates. The best workers request transfers. In this case, fear drives behavior, not excellence.

Toyota, an organization known for its quality, discovers the same door panel issue, but their response is radically different. The worker who spots the misalignment pulls the andon cord; not to report an error, but to signal an opportunity for improvement. The team leader thanks them for catching it. Together, they examine the process and develop a solution to the opportunity.

Here's where Toyota's genius shows: they track "quality confirmations" every time a worker catches and fixes a potential issue before it becomes a defect. These aren't errors; they're saves. A digital board shows running tallies of improvements suggested, implemented, and spread. Workers compete to contribute the most *kaizen* (continuous improvement) ideas.

During daily huddles, team leaders share successful improvements, not problems. No mandates, just invitations to excellence. Workers teach each other their techniques which drives pride in helping their colleagues succeed.

Result: Toyota achieves defect rates ten times better than the industry average, not through punishment and surveillance, but by making improvement visible, celebrated, and contagious. Workers submit thousands of improvement ideas yearly because the system rewards spotting opportunities, not hiding problems. Excellence spreads naturally when you feed success instead of fighting failure.

Types of Feedback Loops in Atomic Systems

Positive Reinforcement Loops:

These amplify beneficial actions:

Success → Recognition → Motivation → More success

When people see their wins acknowledged, they push harder. Small victories compound into major achievements.

Innovation → Reward → Experimentation → More innovation

Rewarding new ideas creates safety for risk-taking. One breakthrough leads to ten more attempts.

Collaboration → Results → Trust → More collaboration

Teams that win together work together more. Trust builds through shared success, not team-building exercises.

Negative Dampening Loops:

These naturally reduce harmful actions:

Waste → Visibility → Embarrassment → Reduction

Nobody wants to be the department throwing away the most product. Make waste visible, and watch it shrink.

Errors → Immediate consequences → Learning → Prevention

Quick, natural consequences teach better than delayed performance reviews. The faster the feedback, the faster the learning.

Conflicts → Quick resolution → Understanding → Harmony

Address friction immediately and publicly. Others learn what to avoid. Team cohesion grows through resolved conflicts, not avoided ones.

Learning Acceleration Loops:
These speed up capability building:

Attempt → Feedback → Adjustment → Mastery
Rapid cycles beat perfect planning. Try, learn, adjust, and repeat. Mastery comes through iteration, not instruction.

Teaching → Questions → Deeper understanding → Better teaching
The teacher learns more than the student. Questions reveal gaps. Understanding deepens through explanation.

Experimentation → Data → Insights → Better experiments
Each test improves the next one. Data drives decisions. Insights compound into breakthrough innovations.

Alignment with Organizational Goals

Effective systems connect individual actions with overarching organizational objectives, but most alignment efforts feel forced, corporate, and even disconnected from daily work. This is the paradox of the world of work today. Often the decision makers are disconnected from reality, not because they choose to be, but because their goals are different. After all, isn't a system built to achieve goals? As Deming said, "A system must have an aim. Without an aim, there is no system".

Through this very thought, atomic alignment is different. There is no paradox. It makes the connection natural, visible, and energizing. As I like to say, everyone is rowing at the same pace.

Traditional Alignment:

- Cascading annual goals
- Quarterly check-ins
- Individual metrics tied to bonuses (go back to the Red Bead Experiment)
- Feels like additional work

This creates distance between action and purpose. People work for metrics, not meaning. Goals feel imposed, not owned. Energy drains instead of builds, creating this world of monotony. Have we forgotten what our intent of work is or do we just chase the metric? This is a question I often ask myself, nearly once a week.

Atomic Alignment:

- Daily actions visibly impacting team goals
- Real-time progress visibility
- Collective wins celebrated immediately
- Feels like meaningful work

This creates an immediate connection between effort and outcome. Progress becomes tangible, success becomes shared, and work gains meaning through visible impact. This is the very thing of the Toyota example mentioned earlier. Let's call this *atomic transparency*.

The Action-Guiding Framework

Based on my research and implementation experience, here's a framework for guiding atomic actions through systems:

Level 1: Make It Obvious

Remove all friction between intention and action. If someone wants to do the right thing, the environment should make it effortless. Visual cues, tool placement, default settings; everything points toward the atomic action.

Level 2: Make It Social

Transform individual actions into shared experiences. People don't just do the action; they see others doing it, share successes, and build momentum together. Isolation kills atomic actions; connection multiplies them.

Level 3: Make It Automatic

Stop relying on motivation and memory. Embed atomic actions into the natural flow of work so they happen by default. The system assumes the action will occur and builds around that assumption.

Level 4: Make It Adaptive

Build in evolution from day one. The system watches what works, spreads successful variations, and kills what doesn't - not through committees or annual reviews, but through continuous natural selection.

Let's imagine how these principles might transform a struggling distribution center with 200 workers facing accuracy, speed, and morale challenges.

The Challenge: Traditional solutions (more training, stricter oversight, threats, discipline) weren't working.

Reminder: Chain Reaction Coefficient Formula (CRC) = (Primary Benefits + Secondary Benefits × 0.7) × Network Reach / (Direct Investment × 1.5)

The Atomic System Design:
Morning Activation (Level 1: Make It Obvious)

- Simple five-minute huddle to start each shift
- Visual boards showing yesterday's wins and today's focus
- Hypothetical Investment: ~$333/day (5 min × 200 workers × $20/hr)
- Potential Impact: Fewer early-shift errors, improved team alignment
- Estimated CRC: Could reach 7-8:1 through error prevention alone

Pick Buddy System (Level 2: Make It Social)

- **E**xperienced workers paired with newer ones
- Partners checking each other's work and sharing techniques
- Hypothetical Investment: ~$150/day for coordination
- Potential Impact: Knowledge transfer, error catching, and relationship building
- Estimated CRC: Could achieve 10-12:1 through combined benefits

Error Celebration (Level 3: Make It Automatic)

- Rewards for finding problems before they ship
- Error-spotting built into the process, not inspection after
- Hypothetical Investment: ~$50/day for tracking system
- Potential Impact: Culture shift from hiding to finding issues
- Estimated CRC: Could reach 12-15:1 by preventing shipped errors

Innovation Hour (Level 4: Make It Adaptive)

- Weekly time for workers to test improvements
- System tracking and spreading successful innovations
- Hypothetical Investment: ~$4,000/week
- Potential Impact: Continuous process improvement, ownership mindset
- Estimated CRC: Could achieve 5-6:1 through accumulated improvements
- Energy Restoration (Supporting All Levels)
- Strategic micro-breaks to maintain peak performance
- Hypothetical Investment: ~$667/day
- Potential Impact: Sustained energy, fewer errors, and better retention
- Estimated CRC: Could reach 8-10:1 through sustained performance

The Integrated Guidance System:

- Digital boards making progress visible (Obvious)
- Peer coaches spreading best practices (Social)
- Color-coded paths building habits (Automatic)
- Weekly innovation sharing sessions (Adaptive)

Hypothetical Outcomes: If implemented effectively, such a system might achieve the following:

- Meaningful accuracy improvements (moving from "good" to "excellent")
- Increased throughput from better flow and fewer errors
- Reduced turnover as work becomes more engaging

■ Higher satisfaction from autonomy and recognition

The Key Insight: Notice how modest investments (~$1,200/day across all initiatives) could theoretically generate value many times greater. The highest-impact actions (like Error Celebration) often require the smallest financial investment because they leverage human nature rather than fighting it. Each element reinforces the others, creating potential for compound returns far exceeding the sum of individual parts. Interestingly, you probably already know this, however, much like your atomic actions, results such as these are invisible.

Common Pitfalls in System Guidance

Here's the sobering truth: for every distribution center success story, there are ten failures gathering dust in corporate graveyards somewhere. The difference isn't in the quality of the atomic actions or the enthusiasm at launch. It's in the subtle mistakes that slowly strangle momentum. Outside of the hypotheticals and simulations, I've watched brilliant atomic systems die not from opposition but from well-meaning missteps that could have been avoided.

Pitfall 1: Over-Engineering
Systems should guide, not control. Too many rules, steps, or requirements kill atomic energy.

Pitfall 2: Ignoring Informal Systems
Every organization has formal systems (policies) and informal systems (how things really work). Atomic actions must align with both.

Pitfall 3: One–Size–Fits–All
Different teams need different guidance. What works in accounting might fail in sales.

Pitfall 4: Set and Forget
Systems need constant gardening. Without attention, even the best systems decay.

The distribution center example succeeded because they avoided every one of these pitfalls. Their system was simple, not over engineered. They worked with existing relationships, not against them. Each department adapted the approach to their reality, and they kept refining, never assuming they were done. This is critical. Adapting to reality. Because let's face it, as we grow in leadership, we are always told "don't forget where you came from" but the truth is: we all do; we all forget where we came from.

Your System Challenge

This week, choose one atomic action, and build a guidance system around it:

Create an environmental cue (visual reminder, tool placement, etc.).

Add a feedback mechanism (simple tracking, visibility, etc.).

Include a social element (partner, team goal, sharing, etc.).

Design in evolution (how will it improve over time?).

Measure the multiplied impact.

Remember: atomic actions without systems are just good intentions. Systems without atomic actions are just bureaucracy, but atomic actions within guiding systems is transformation.

The future belongs to organizations that can orchestrate thousands of small actions into symphonies of productivity. Will yours be one of them?

Guiding atomic actions through systems creates tremendous power, but power without control creates destruction, not productivity. That's why every nuclear reactor has control rods. Your organization needs them, too.

CHAPTER 9

The Multiplication Effect

When atomic actions reach critical mass

There's a moment in every transformation when something magical happens. The atomic actions you've carefully planted and nurtured suddenly begin multiplying on their own. What started as individual efforts became a movement. What began as isolated improvements became systemic change. This is the multiplication effect, and understanding it is crucial to sustaining your atomic transformation.

The Tipping Point of Transformation

Malcolm Gladwell wrote about tipping points, those dramatic moments when small changes make a big difference. In atomic productivity, I've discovered the tipping point occurs predictably when 15–20 percent of daily actions become atomic (CRC > 5:1).

At this threshold, something shifts. People stop asking, "Why should we?" and start asking "Why haven't we?" Innovation becomes expected, not exceptional. Improvement becomes automatic, not forced.

You'll know when your organization approaches critical mass. After months of implementing atomic actions with steady but modest results, something will shift. Ideas will multiply faster than you can track. Teams will start collaborating without prompting, and metrics will improve mysteriously. What's happening? You've reached critical mass.

Understanding Multiplication Mechanics

To harness the multiplication effect, you need to understand how it works. Unlike simple addition where 1+1=2, multiplication in human systems follows different rules.

The Network Effect shows that each person influenced by an atomic action becomes a potential multiplier. When one person adopts a high CRC practice, they typically influence three to five others. Those others influence three to five more. Within weeks, a single atomic action can touch hundreds. The mathematics are compelling: one person reaching five others, who each reach five more, touches 155 people in just three degrees of separation. That's the power of network multiplication.

The Variation Advantage demonstrates that as atomic actions spread, they don't replicate exactly but instead evolve. Each person adds their own improvements, creating variations that often work better than the original. This beneficial mutation accelerates improvement. Think of it as organizational evolution in fast forward: the best adaptations survive and spread, creating a diverse ecosystem of solutions optimized for different contexts.

The Compound Learning Effect shows that each atomic action makes the next one easier to implement. People develop what I call "atomic intuition," which is the ability to spot high CRC opportunities naturally. Organizations build "atomic muscle memory," meaning the capability to implement changes quickly. What takes months initially takes weeks, then days. The learning curve doesn't just improve; it shortens exponentially.

These three mechanisms work together: network spread, beneficial variation, and accelerating capability create multiplication that far exceeds simple arithmetic progression.

Catalyzing Multiplication

While multiplication happens naturally once you reach critical mass, you can accelerate it through deliberate strategies.

Create Multiplication Mechanisms that make spreading atomic actions easier than keeping them secret.

Build Story Engines that turn dry improvements into compelling narratives.

Design Recognition Systems that celebrate both creators and adopters of atomic actions.

Establish Learning Loops that capture what works and what doesn't.

These strategies can accelerate the spread of positive atomic actions throughout your organization, but here's the critical insight:

the same mechanisms that multiply beneficial actions can also multiply harmful ones. The very power that makes atomic productivity transformational also makes it potentially dangerous. Just as nuclear reactions can power cities or destroy them, multiplication effects in organizations can build or burn.

This brings us to an essential warning...

The Dark Side of Multiplication

Not everything that multiplies is beneficial. I've seen the multiplication effect worked against organizations.

Toxic Multiplication occurs when low-CRC or negative actions spread. Let's assume a company accidentally created a culture of "metric gaming" when they over-rewarded a poorly designed atomic action. The gaming behavior multiplies faster than they could contain it. It would take months to root out.

What's the antidote? Vigilant monitoring of what's actually multiplying, not just what's intended to multiply. Track both formal and informal spread. Observe what people are doing, not just what they're reporting. This is the toughest part. We become obsessed with reporting out, often in a manner that benefits us, not the action.

Multiplication Fatigue happens when too many atomic actions multiply simultaneously. People become overwhelmed by constant change, even positive change. Energy depletes. Cynicism creeps in, and progress stalls.

What's the antidote? This is exactly why you need control rods (Chapter 7). Before implementing formal controls, you can prevent fatigue by celebrating stability periods between multiplication waves, ensuring people have time to digest and integrate changes, and maintaining core routines that provide psychological anchors.

Innovation Inflation occurs when the bar for "atomic action" drops too low. People start claiming every minor task completion as an innovation. The currency of improvement becomes devalued. True atomic actions get lost in the noise.

What's the antidote? Maintain standards for what constitutes an atomic action (CRC > 5:1), regularly recalibrate expectations, celebrate quality over quantity, and teach the difference between improvement and true multiplication.

Measuring Multiplication

How do you know if multiplication is really happening? Traditional metrics won't tell you. You need multiplication metrics.

The Spread Rate tracks how quickly atomic actions propagate through your organization. Calculate: (Number of people using action) / (Weeks since introduction). A healthy spread rate exceeds ten people/week for high-impact actions.

The Variation Index measures how many beneficial variations emerge from each original atomic action. Calculate: (Number of documented variations) / (Number of original actions). Healthy organizations show three to five variations per original action.

The Innovation Velocity tracks how quickly your organization goes from idea to implementation. Measure the average time from suggestion to full adoption. In multiplying organizations, this shrinks from months to weeks to days.

The Cultural Saturation Score estimates what percentage of daily actions have become atomic. Survey random samples of work, calculating what portion shows CRC > 5:1. Remember, 15–20 percent is the tipping point for multiplication.

Your Multiplication Checklist

To create multiplication in your organization, ensure you have the following:

- ☐ Clear examples of successful atomic actions (proof it works)
- ☐ Easy mechanisms for sharing improvements (low friction spread)
- ☐ Recognition for both creators and adopters (incentive alignment)
- ☐ Safe spaces to experiment and fail (psychological safety)
- ☐ Time allocated for trying new approaches (resource commitment)
- ☐ Visible tracking of multiplication effects (measurement systems)
- ☐ Stories that inspire rather than mandate (emotional connection)
- ☐ Leaders who model atomic thinking (top-down support)
- ☐ Protection from multiplication fatigue (sustainable pace)
- ☐ Standards for what constitutes true atomic action (quality control)

The Multiplication Mindset

The deepest change isn't in the actions themselves but in how people think. When multiplication takes hold, employees stop seeing themselves as task-doers and start seeing themselves as system improvers. They stop asking, "What's my job?" and start asking,

"What could work better?" They stop hoarding improvements and start sharing freely.

This mindset shift is the real prize. When everyone thinks atomically, looks for multiplication opportunities, and shares improvements freely, that's when organizations become truly unstoppable.

PART IV

Measurement and Technology

CHAPTER 10

Metrics That Matter

Moving beyond vanity metrics to transformational measurement

"Without data, you're just another person with an opinion," Deming said. Here's what he didn't say: with the wrong data, you're worse off than having no data at all. I was once listening to a podcast episode featuring the CTO (Shyam Sankar) of Palantir, a software company that helps organizations integrate data, decisions, and operations. He mentioned that there is nothing inherently valuable about data. It only becomes valuable when used to make informed decisions.

The Metrics Trap

Most organizations fall into what I call the metrics trap. They measure what's easy instead of what's important, track activities instead of outcomes, focus on individual metrics instead of system metrics, use metrics for punishment instead of improvement, and collect data without action plans.

In my grocery retail days, we tracked items scanned per minute, which made cashiers rush and make errors. We measured bathroom cleaning frequency but not cleanliness. We tracked the number of "holes" on a shelf but not the effectiveness of the product, among many other things. In this case, we were optimizing for metrics, not performance. The metrics had become the goal, not a tool for reaching goals.

Key Performance Indicators vs. Key Progress Indicators

Traditional KPIs often measure lagging indicators, things that already happened. For atomic productivity, we need Key Progress Indicators (KPgIs), metrics that show momentum, not just position.

Traditional KPIs include last month's sales, quarterly productivity, annual turnover rate, and customer satisfaction score. These tell you where you've been, not where you're going. Atomic KPgIs, on the other hand, track daily chain reactions started, CRC of implemented actions, system stress indicators, innovation velocity, and energy multiplication rate. The difference? KPgIs help you steer, not just report.

The Meta-Metric

The most important metric in any atomic system? The Metric Effectiveness Score (MES):

MES = (Decisions improved + Behaviors aligned + Innovations sparked) / (Cost to collect + Time to analyze + Confusion created)

If your MES is below 1, your metrics are hurting more than helping. For instance, if your daily sales report takes two hours to compile and reaches twenty managers, but only two managers use it for decisions while eighteen find it confusing, your MES would be: $(2 + 0 + 0) / (2 \text{ hours} \times 20 \text{ people} + 18 \text{ confusion points}) = 0.04$. Time to kill that report.

The Meta-Metric reveals a truth: most organizations are data-rich but insight-poor. To reverse this, you need metrics designed specifically to surface and multiply atomic actions. Traditional metrics won't cut it.

The Atomic Metrics Framework

Through my research and simulations, I've developed a framework for metrics that actually drive atomic productivity.

Level 1 focuses on System Health Metrics. These tell you if your system can support atomic actions. You need to track change absorption capacity, feedback loop speed, innovation suggestion rate, cross-functional collaboration frequency, and system stress indicators. Without a healthy system, atomic actions can't take root.

Level 2 encompasses Chain Reaction Metrics. These measure multiplication effects by tracking average CRC of daily actions, secondary benefits generated, ripple effect distance, compound improvement rate, and sustainability scores. These metrics show you whether your atomic actions are creating true chain reactions or just isolated improvements.

Level 3 involves Human Energy Metrics. These track the fuel for atomic productivity through engagement pulse scores, initiative participation rates, peer teaching frequency, celebration occurrence, and burnout early warnings. Remember, humans are the atoms in your productivity chain reaction. If their energy depletes, the reaction stops.

Level 4 covers Outcome Acceleration Metrics. These show whether atomic actions are achieving goals by measuring time to goal achievement, resource efficiency improvements, quality gains per effort unit, innovation implementation speed, and customer impact velocity. These metrics connect atomic actions to business results.

Statistical Process Control in the Atomic Age

Shewhart and Deming gave us Statistical Process Control (SPC), the ability to distinguish between common cause and special cause variation. Most organizations apply SPC to the wrong things.

Traditional SPC application tracks defect rates, sets control limits, investigates when limits are exceeded, and tries to eliminate all variation. This approach assumes variation is bad and stability is good.

Atomic SPC application is different. It tracks improvement rates, sets growth expectations, investigates when improvement stalls, and cultivates beneficial variation. This approach recognizes that some variation signals innovation and improvement.

Here is a meaningful example. A distribution center used SPC on their pick rates. When someone performed "too well," they were investigated for cheating. This killed innovation. We redesign their approach to track improvement ideas generated, set lower control limits for minimum innovation expected, investigate when teams aren't experimenting, and celebrate beneficial variations. As a result continuous improvement became the norm, not the exception.

Data-Driven Decision Making That Actually Works

Here's the brutal truth: most "data-driven" decisions aren't truly data driven. They're opinion-driven decisions with data decorations. Referring back to Shyam's position on data, another notable statement he made was that people think data is the new oil; however, he contradicts that thought with the notion that data is the new 'snake oil,' True data-driven decision-making in atomic systems looks different from traditional thought.

First, define decision criteria before collecting data. What will you do if the data shows X? If you don't know, don't collect it. Too many organizations collect data hoping it will tell them what to do. Often enough, and I've seen this countless times, many 'practitioners' in the workplace seem to force data as well, forcing it to tell us what isn't there.

Second, measure actions, not just outcomes. Outcomes lag while actions lead. Track both, but steer by actions. If you're only measuring results, you're driving by looking in the rearview mirror.

Third, create feedback loops, not reports. Data should flow to decision-makers in real time, not quarterly presentations or weekly report out meetings. By the time a quarterly report or weekly meeting is compiled, analyzed, and presented, the opportunity for intervention has passed.

Fourth, build in response triggers. When a metric hits a threshold, what happens automatically? Design it in. Don't wait for someone to notice and decide. Make the response part of the system.

Fifth, measure the metrics. Are your metrics driving the right behaviors? That's the most important metric of all. I've seen metrics that technically improved while actual performance declined because people gamed the system. Take this into account. When someone games the system, they've at least figured out how the system works. Use this to your advantage.

Balancing Quantitative and Qualitative Metrics

Numbers tell you *what*. Stories tell you *why*. You need both. Most organizations live in a quantitative paradise that masks a qualitative nightmare. The dashboards glow green with met targets and exceeded benchmarks. Resolution rates hit their marks. Satisfaction scores meet minimum thresholds. By every spreadsheet measure, there is success, but spreadsheets don't capture the full truth. They can't measure the exhaustion in a voice, the creativity being suffocated by scripts, or the resignation building behind forced smiles. They don't show when efficiency has become so ruthless that humanity has been optimized out of the equation.

I can admit... I love data, spreadsheets, and any tool that helps drive the quantitative aspect of work, but I don't passionately care about numbers. I don't love the goals and targets because they lead to a point of content. They lead to a realm of comfortability that stifles continuous improvement and growth.

The path forward isn't choosing between numbers and narratives. It's weaving them together into a complete picture. Keep your efficiency metrics, but add meaning metrics. Track resolution rates alongside relationship depth. Measure speed, but also measure the energy left in the interaction. Include the voices of those delivering the service, not just those receiving it.

When you balance the quantitative with the qualitative, the real story emerges. Often, what looks like success is actually a system running on borrowed time, efficient but brittle, meeting targets while building toward collapse. Only when you see both dimensions can you design atomic actions that transform not just the numbers, but the human experience that drives sustainable performance.

True atomic productivity lives at the intersection where metrics meet meaning.

The Acceptability vs. Desirability Framework

I first learned of this concept from Dr. Bill Bellows, formerly the Deputy Director of the Deming Institute and a good friend and mentor. It transformed how I think about metrics.

Most metrics define acceptability. Did we meet minimum standards? Are we within tolerance? Did we hit our targets? Acceptability isn't excellence. As Bill helped me understand with the example of doctors graduating from med school, all doctors are acceptable (they graduated). Would you want a doctor who graduated last to be your primary physician? You'd likely desire the doctors who finished top of their class. This distinction reveals a profound truth about how we measure performance. Acceptability asks, "Is it good enough?" Desirability asks, "Is it remarkable?" The gap between these questions is where atomic transformation lives.

Consider how this plays out across industries. An acceptable flight gets you there safely. A desirable flight makes you eager to fly that airline again. An acceptable meal meets health codes. A desirable meal creates memories. (Think Olive Garden vs. your favorite local Italian bistro.) An acceptable employee meets their quotas. A desirable employee transforms everyone around them.

The shift from acceptability to desirability metrics changes everything. Instead of asking, "Did we avoid complaints?" we ask, "Did we create advocates?" Instead of measuring deadlines, we track expectations. Instead of counting "Zero defects," we celebrate "Moments of delight."

Atomic metrics should drive toward desirability. Are we delighting, not just satisfying? Are we improving, not just maintaining? Are we innovating, not just operating?

When organizations make this shift, the multiplication effect is extraordinary. Teams stop optimizing for minimum standards and start competing to create excellence. The energy changes from compliance to creativity. The CRC of moving from acceptable to desirable is often unmeasurable because you've transformed not just performance, but possibility.

The Role of Variation in Metrics

Deming emphasized that understanding variation is critical to improving productivity, but most organizations try to eliminate variation instead of understanding it.

In atomic systems, variation is information. Consistent performance might mean missed opportunities. Beneficial spikes show what's possible. Problematic dips reveal system failures. Patterns tell stories.

When organizations see variation in performance metrics, the instinct is to standardize. Force everyone to the mean. Create rigid procedures. Eliminate the outliers. This fundamentally misunderstands what variation is telling you.

Often, that 30 percent variation between workers isn't a problem to solve; it's intelligence to decode. The high performers aren't lucky or working harder; they've discovered something. They've adapted their methods to specific conditions. They've innovated solutions the system designers never imagined. The variation maps to real differences in context, product types, or situational demands that standardization would ignore.

The atomic approach is counterintuitive: instead of eliminating variation, study it. What creates beneficial variation? Which innovations are hiding in those outliers? How can flexibility itself become a strength? When you allow method variation within outcome consistency, something remarkable happens. Performance improves not despite the variation, but because of it.

The key shift is measuring improvement rate alongside absolute performance. Are people getting better? Are innovations spreading naturally? Is the variation teaching you about your system? When you track these dynamics, you often find that maintaining variation while spreading best practices yields far better results than forced standardization ever could.

This is the paradox of variation in atomic systems: sometimes the messiness is the message. Sometimes the differences are the data. Sometimes allowing variation is the most powerful standardization of all, standardizing on continuous improvement rather than rigid compliance. There may be something to consider. Adopting, understanding, and accepting variation does not mean everyone should do something different because they are unique. There is still a difference here. Knowing the difference is true atomic productivity.

Building Your Atomic Metrics System

Here's a practical approach to implementing metrics that matter.

Week 1 involves a Metrics Audit. List every metric you currently track. Mark which ones drive decisions. Note which ones drive behavior (good or bad). Calculate the cost of collecting each. Eliminate metrics that fail the ROI test.

Week 2 focuses on Atomic Metric Design. Identify your key atomic actions. Design metrics that track multiplication. Include both leading and lagging indicators. Balance quantitative and qualitative. Focus on progress, not just position.

Week 3 covers Implementation. Start simple (three to five metrics max). Make them visible. Train people on what they mean. Create response protocols. Begin collecting baseline data.

Week 4 is about Calibration. Review what behaviors metrics are driving. Adjust definitions or thresholds. Add context or storytelling. Ensure metrics serve strategy. Plan first improvements.

Common Metrics Mistakes in Atomic Systems

Most teams approach atomic system metrics like they're measuring individual gears instead of the clockwork, tracking every micro-action while missing the emergent behaviors that actually drive outcomes. Here are four measurement traps that derail even experienced teams, and why fixing them requires rethinking what you measure, not just how you measure.

Mistake 1: Measuring Individual Atomic Actions. Atomic actions work in systems. Measure system effects, not isolated actions.

Mistake 2: Over–Precision. CRC of 7.2 vs 7.3 doesn't matter. Rough accuracy beats false precision.

Mistake 3: Gaming Prevention Focus. If people game your metrics, your metrics are wrong, not your people.

Mistake 4: Set and Forget. Metrics need evolution. What matters in startup phase differs from scaling phase.

Your Metrics Challenge

This week, revolutionize one metric. Choose a current metric that isn't driving improvement. Redesign it to track multiplication, not just activity. Add a qualitative component, and create a response protocol. Watch how behavior(s) shifts.

Remember: metrics should be servants, not masters. They should illuminate atomic actions, not obscure them. They should inspire improvement, not enforce compliance.

In the end, what you measure is what you multiply. Make sure you're measuring what matters.

CHAPTER 11

Technology as a Force Multiplier

Amplifying human capability without losing the human touch

Technology insertion in productivity systems is like nuclear fuel in a reactor. Used correctly, it powers transformation. Used poorly, and it creates expensive disasters.

I once watched two grocery stores in the same region implement identical AI-driven inventory systems. The first store used it as a decision-support tool, giving experienced staff data-backed insights about buying patterns while preserving their ability to adjust for local knowledge. Productivity rose 40 percent. The second store went fully automated, removing human oversight from ordering decisions. Productivity fell 20 percent as the system repeatedly misordered products, unable to account for construction road closures, local school schedules, or seasonal community events. What was the difference? One amplified human capability; the other tried to replace it.

The Technology Insertion Framework

Technology insertion – determining which technologies should be replaced, enhanced, or changed for organizational optimization, requires systematic thinking. Most organizations approach it backwards.

The traditional approach starts with seeing cool technology, then buying technology, forcing people to use it, and finally wondering why it failed. Think of the world of artificial intelligence. Everyone is looking for some sort of AI insertion, not because they see the benefits but because it is what the market or competitive landscape say you must have. The atomic approach is different. It begins with identifying human capability gaps, then finding technology that amplifies human strengths, designing implementation to create chain reactions, and finally measuring multiplication effects.

The Grocery Tech Revolution

My experience in grocery retail and my dissertation research coincided with massive technology adoption in grocery retail. I watched stores implement mobile inventory scanners, automated ordering systems, customer apps, employee scheduling technologies, and robotic floor cleaners.

The success rate was about 30 percent, but the successes were transformational. What made the difference?

Let me share a successful implementation example: Mobile Inventory System. The technology consisted of handheld devices showing real-time inventory, optimal pick paths, and predictive ordering suggestions.

Why it succeeded becomes clear when you examine the approach. It amplified worker knowledge: workers knew what sold, while the system knew what was where (usually). It reduced friction by eliminating the need to walk back to check stock. It created time for customer service. It enabled peer learning as workers could share techniques digitally. It generated data for continuous improvement.

What was the CRC here? 15:1. It was like having a digital assistant, helping you answer customer questions and move faster.

Now consider a failed implementation example: self–checkout systems. While not entirely failed since the usage of self-checkout is very prominent today, I have witnessed the failure of a store fully eliminating human-based cash registers for a full line of SCO robots.

Why it failed is equally instructive. It replaced human connection, making customers feel dismissed. It increased friction as complex issues always occurred, and human interaction was rarely around to help solve problems. It demoralized employees who felt their jobs were threatened. It created negative chain reactions of angry customers and stressed employees. It generated complaints, not solutions. While much of this is likely due to the change factor, the CRC was 0.3:1. Customers said, "I just want to talk to a human who understands."

The Technology Acceptance Model in Practice

Fred Davis's Technology Acceptance Model (TAM) shows that acceptance depends on perceived usefulness and perceived ease of use.

In atomic systems, we need to add perceived amplification (Does it make me more capable?) and perceived humanity (Does it preserve human connection?). These four factors, usefulness, ease, amplification, and humanity, determine whether technology creates multiplication or resistance.

In my dissertation, the TAM was the second framework I used to guide my research along with Deming's System of Profound Knowledge. Using this framework allowed me to consider technology through the lens of organizational investment. If a company decides to insert new technology into its daily operations, whether customer facing or worker facing, it must do so with the intent that it will be used…effectively!

As I continued to research and align with the TAM, experiencing new technology insertion in my day job and even as an adjunct professor, I saw that the preservation of human connection and capability amplifications were minimal or missing entirely. Why? Because there is a seemingly proverbial purpose of saving money, synergizing, or even eliminating jobs for the sake of the latter. This shortsighted approach misses the exponential returns possible when technology amplifies rather than replaces human capability. The organizations that understand this distinction are the ones achieving true atomic productivity.

Building Technology Chain Reactions

The best technology implementations create their own atomic actions. When technology truly amplifies human capability, it doesn't just improve existing processes; it enables entirely new possibilities.

Workers discover optimal paths and methods, sharing them naturally. Innovation emerges from use, not from design. The technology becomes a platform for continuous improvement, generating chain reactions that multiply far beyond the initial investment.

Your Technology Audit

This week, audit one technology in your organization. Ask these simple questions: Does it make humans more capable (amplification) or replace them? Does it reduce friction or add complexity? Find one way to increase amplification then measure the CRC of the improvement. In Chapter 12, you will learn how to use what I call a Technology Success Score.

The Ultimate Technology Test

Here's my simple test for any technology investment:

After implementation, do humans feel more capable or less capable?

If the answer is "less capable," you're not implementing technology. You're implementing expensive friction.

In atomic productivity, technology is nuclear fuel. It should power human chain reactions, not replace them. The goal isn't a fully automated system; it's a system where humans and technology together create exponential value.

Because in the end, the best technology disappears into the background, quietly multiplying every human action into extraordinary impact.

After all this discussion of technology as a force multiplier, you might be tempted to automate everything, to chase the promise of infinite CRC through perfect technical systems. That would miss the most crucial lesson from my research.

I've watched organizations intoxicated by technology's promise crash into a brutal reality: even the most sophisticated systems fail without human judgment, creativity, and adaptation. Conversely, I've seen simple technologies create extraordinary transformations when they honor and amplify human capability.

This isn't a paradox; it's a design principle. The highest CRC doesn't come from replacing humans or from avoiding technology. It comes from finding the sweet spot where each amplifies the other's strengths.

What you're about to read isn't a retreat from technology advocacy. It's the completion of the picture. Atomic productivity isn't achieved through human effort alone or technological sophistication alone. It's achieved when we create systems where humans and technology dance together, each making the other more capable than they could ever be alone.

The future doesn't belong to organizations that choose humans over machines or machines over humans. It belongs to those who render the choice obsolete by creating something greater than either could achieve alone.

That's the balance we must strike. That's the multiplier effect we must capture. That's the atomic potential waiting to be released.

CHAPTER 12

The Human–Technology Balance

*Creating systems where humans
and machines amplify each other*

"Are robots going to take our jobs?"

During my dissertation research, I heard this question countless times from grocery workers watching self-checkout expand, inventory robots roll through aisles, and AI systems make decisions that used to require human judgment. Even today, as I write this, I cross the path of many people who think AI will take their jobs.

They were asking the wrong question. The real question is, "How can humans and technology multiply each other's capabilities?"

The False Binary

We've been conditioned to see human vs. machine as a binary choice. This thinking creates fear-based resistance to beneficial technology, dehumanizing automation that destroys value, missed opportunities for multiplication, and adversarial relationships instead of partnerships.

The atomic productivity approach rejects this binary. Instead, we see a spectrum of human-technology integration where the goal is finding the optimal mix for exponential impact, not simply assuming that we can fix organizational EBIT(DA) opportunities by replacing humans with technology.

The Human-Technology Balance Formula

Through my computational simulations, research, and field experience, I developed this formula:

Technology Success = (Human Amplification × System Integration) / (Friction Added + Human Replacement)

Let's break this down. Human Amplification asks whether technology makes humans more capable. System Integration examines whether it works with existing workflows. Friction Added measures how much complexity it introduces. Human Replacement calculates how much human connection it eliminates.

The best technology scores high on the numerator, low on the denominator.

To use this formula and properly calculate your score, you will score each element from 0 to 10.

Human Amplification: 0 = humans can do less than before; 10 = humans can do exponentially more. Ask: How many new capabilities does this enable?

148

System Integration: 0 = requires complete process overhaul; 10 = seamlessly fits current workflows. Ask: How many existing processes must change?

Friction Added: 0 = no new complexity; 10 = significant training, steps, or confusion. Ask: How much harder is the work now?

Human Replacement: 0 = no human interaction lost; 10 = completely eliminates human elements. Ask: What human connections disappear?

Example: A mobile inventory scanner might score: Human Amplification (8) × System Integration (7) = 56 in the numerator. Friction Added (2) + Human Replacement (1) = 3 in the denominator. Technology Success Score = 56/3 = 18.7. Anything above 10 indicates strong atomic potential.

The Multiplication Matrix

Through my research, I developed a framework for understanding the human-technology balance.

Quadrant 1 encompasses Human Only (Low Tech, High Touch). Examples include therapy, creative strategy, and relationship building. The strengths are empathy, judgment, and innovation. The weaknesses are scalability, consistency, and fatigue. The atomic potential is low, limited by human constraints.

Quadrant 2 represents Human Amplified (Tech-Enabled Humans). Examples include surgeons with robotic assistance and analysts with AI tools. The strengths are enhanced capability, reduced errors, and increased reach. The weaknesses are training requirements and technology dependence. The atomic potential is very high, combining human judgment with machine power.

Quadrant 3 involves Machine Amplified (Human-Guided Automation). Examples include AI with human oversight and automated systems with exception handling. The strengths are scale, speed, and consistency. The weaknesses are rigid thinking and context blindness. The atomic potential is high, merging machine efficiency with human wisdom.

Quadrant 4 contains Machine Only (Full Automation). Examples include data center operations and repetitive manufacturing. The strengths are perfect consistency, 24/7 operation, and cost efficiency. The weaknesses are brittleness, no adaptation, and no innovation. The atomic potential is low, lacking the human spark for chain reactions.

The sweet spots? Quadrants 2 and 3, where humans and technology multiply each other.

Quadrant 1 Human Only	Quadrant 2 Human Amplified ★
■ Therapy ■ Creative strategy ■ Relationship Building **Atomic Potential: Low**	■ Surgeon w/robots ■ Analysts w/AI **Atomic Potential: Very High**
Quadrant 3 Machine Amplified ★	Quadrant 4 Machine Only
■ AI w/human oversight ■ Automated + exception ■ Human guided systems **Atomic Potential: High**	■ Data centers ■ Full automation **Atomic Potential: Low**

Real-World Balance Examples

The Deli Counter Revolution illustrates this balance perfectly. The traditional approach involved customers waiting while an employee takes the order, slices meat, wraps, weighs, prices, and hands it over. The tech-only approach would be a vending machine with pre-sliced meat - efficient but soulless. Let's face it. We'd probably all question when that hunk of meat was loaded into the machine anyway.

The atomic balance approach combines both worlds. A digital ordering kiosk handles standard orders while a human specialist manages custom requests and recommendations. Automated slicing operates with human quality control. Technology handles the transaction while humans handle the connection - a win-win for the world that we are all living in.

The Trust Equation

For human-technology balance to work, trust must flow both directions.

Humans Trusting Technology requires reliability (Does it work consistently?), transparency (Can I understand its decisions?), override ability (Can I intervene when needed?), and amplification (Does it make me better?).

Technology Trusting Humans involves appropriate use (Do humans use it as designed?), feedback integration (Do humans help it improve?), exception handling (Do humans cover its blind spots?), and partnership mindset (Do humans see it as a tool, not a threat?).

Building Balanced Systems

Here's my approach to creating human-technology balance.

Step 1: Task Decomposition. Break down work into components. What requires human judgment? What requires human connection? What requires scale or speed? What requires perfect consistency?

Step 2: Optimal Assignment. Assign each component to its strength. Humans excel at judgment, creativity, empathy, and adaptation. Technology excels at calculation, memory, repetition, and scale.

Step 3: Interface Design. Create seamless handoffs with clear transition points, shared information displays, override mechanisms, and feedback loops.

Step 4: Continuous Calibration. The balance point shifts as technology improves, humans adapt, needs change, and systems evolve.

Using these four steps helps you determine the gaps that exist. Atomic productivity isn't a guide to tell you where you need technology or humans, but it can guide you where to look at what to consider for the perfect balance.

The Psychology of Partnership

The biggest barrier to human-technology balance isn't technical; it's psychological. When people feel valued rather than replaced, productivity soars. Yet, most implementations focus on efficiency metrics while ignoring human dignity. Technology must make their work more important, not less.

Consider Deming's insight about human nature: people have an innate need for relationships, self-esteem, and joy in work. Empowerment beats control every time. Instead of restricting capabilities, technology should expand what's possible, opening new avenues for mastery and contribution.

Frame the relationship carefully. Technology as teammate, not the boss. This isn't just semantics; it's fundamental psychology that drives real results. People resist subordination but embrace partnership.

Growth remains non-negotiable. Stagnation kills motivation faster than any technology rollout. Use these tools to enable continuous learning and development. The most successful implementations recognize what Deming understood: respecting human psychology isn't optional. It's the foundation upon which all sustainable improvement builds.

The Evolution Principle

Most organizations design their human-technology balance once and assume it's done. Six months later, they wonder why everything's breaking. What's the truth? Technology capabilities change, human skills develop, business needs shift, and static balance becomes imbalance. That's why you must build evolution into the design from day one. Human-technology balance isn't static. It evolves. The key is to design systems that improve their own balance over time.

Learning Loops ensure technology learns from human decisions, humans learn from technology patterns, and both adapt based on outcomes. When your customer service AI suggests responses, agents rate them. Good suggestions get reinforced; poor ones get corrected. Within months, the AI mirrors your best agents' judgment.

Flexibility Mechanisms make it easy to shift tasks between human and machine, enable quick testing of new balance points, and provide rollback capabilities if balance fails. Start with humans doing 80 percent of a task, technology 20 percent. Built-in weekly reviews let you slide that balance as technology proves itself. Always maintain a 'full manual' option for when technology fails. Tune into your variation. (Remember that variation is music.)

Success Metrics track not just efficiency but effectiveness, not just output but outcome, not just satisfaction but growth. Instead of tracking 'calls per hour,' track 'problems solved completely.' Instead of 'processing time,' measure 'customer effort required.' The metrics themselves drive evolution toward better balance

The Future of Balance

As AI and automation advance, the human-technology balance becomes more critical, not less. The winners won't be organizations that automate everything or resist all technology. They'll be those that find the sweet spot where humans feel more capable because of technology, technology performs better because of human guidance, both evolve together in a virtuous cycle, and chain reactions multiply exponentially.

The Balance Manifesto

Here's what I believe about human-technology balance:

Humans are not resources to optimize but capabilities to amplify. Technology is not a replacement for humans but a multiplier of human potential. The best systems make both humans and technology better. Balance is not a fixed state but a dynamic dance. The goal is not efficiency alone but exponential impact.

In atomic productivity, the most powerful chain reactions come from human creativity sparked by technological capability. Neither alone can match what both together achieve.

The future belongs to organizations that master this balance. Will yours be one of them?

PART V:

The Psychology of Systems

CHAPTER 13

Understanding the Willing Worker

*Unleashing the atomic potential
already in your organization*

In Chapter 2, we saw how Deming's willing workers were defeated by bad systems in the Red Bead Experiment. Now let's explore how this principle transforms modern organizations. Today's workplace is far more complex than Deming's era. When Deming first began advocating for systems thinking and the protection of willing workers, the workplace was dominated by manufacturing, standardized labor, and top-down control. Today's world is far more complex. Work is digital, distributed, and dynamic. Employees are more educated, more mobile, and more purpose driven. Complexity is higher, and the pace of change is exponential, especially with the speed of artificial intelligence.

And yet, the core issue remains the same: good people are still trapped in bad systems. Only now, the systems are often digital, data-driven, and deceptively automated, which means leaders are often even more blind to the obstacles workers face.

We use dashboards, KPIs, and AI tools believing they will drive productivity, but these tools can just as easily dehumanize and demoralize when applied without context or feedback. In today's workplace, Deming's warning is more relevant than ever.

The Systems Barrier to Contribution

Willing workers face predictable barriers in every organization. The patterns are universal: frontline employees see inefficiencies but lack authority to fix them. Support staff know what customers need but must follow rigid scripts. Managers understand what their teams require but drown in administrative obligations. Knowledge workers have innovative solutions but can't navigate approval bureaucracy. These aren't performance problems. They're system failures.

The tragedy compounds over time. After months or years of suggestions being dismissed, willing workers learn a devastating lesson: caring hurts. Initiative gets punished with "that's not your job." Innovation meets "that's not how we do things." Eventually, the brightest, most capable employees make a rational choice: they stop trying, and their engagement declines.

This is how organizations systematically destroy their own atomic potential - not through malice, but through systems that make contribution harder than compliance. Every suppressed idea, every ignored suggestion, every "stay in your lane" response teaches workers that their willingness is unwanted.

The cost is immeasurable, because when willing workers stop trying, you haven't just lost their ideas. You've lost their energy, their innovation, and eventually, them. They either quit and leave, or worse… they quietly quit!

Designing for Willingness

Leaders must shift from motivating individuals to removing systemic friction. This transformation requires several key elements.

Creating Clear, Role-Aligned Workflows is essential. In my experience, I found that roles are not often aligned with job descriptions, and vice-versa. I even teach this in a Human Resources Management course. Job descriptions are not updated to align with the duties of one's job, yet, the very thing we use to hire is the job description.

Encouraging Experimentation Without Fear transforms culture. A manager implementing a program that allows for workers to test new ideas and theories without punishment for challenging the status quo.

Measuring What Matters – Not Just What's Convenient changes everything. Instead of tracking tasks completed, KPIs met, or targets achieved, measure what counts. This might include behaviors not accounted for to achieve target etc. Consider this. Convenient metric breed convenient behaviors.

The Psychology of Unleashing

Here's what happens psychologically when you unleash willing workers.

Stage 1: Skepticism. "This is just another management fad." Workers have been burned before. They wait to see if you're serious.

Stage 2: Testing. "Let me try one small thing." Workers tentatively offer ideas, watching for punishment or dismissal.

Stage 3: Acceleration. "Wow, they actually listened!" Success breeds courage, ideas flow faster, and energy builds.

Stage 4: Ownership. "This is our system now." Workers become fierce protectors and improvers of the system they helped create.

This transformation from skepticism to ownership isn't just a feel-good story; it's the critical path to sustainable productivity gains. When workers own the system, they stop working around it and start working to improve it. The difference between compliance and commitment shows up in every metric that matters.

Small Wins and Engagement

Studies show that the single biggest driver of workplace motivation is progress. The feeling of forward movement. When systems allow willing workers to make and track small improvements, engagement soars. These are the atomic actions that compound into systemic transformation.

The most effective approach is radically simple: create visible spaces where any improvement, no matter how small, can be shared without approval barriers or minimum thresholds. This could be physical boards, digital platforms, or regular sharing sessions. The medium matters less than the principle: make progress visible and celebration immediate.

When organizations implement this approach, predictable patterns emerge. Micro-improvements multiply from dozens to hundreds monthly. Peer learning accelerates as workers discover they're all solving similar problems. Positive competition develops, not for rewards, but for the satisfaction of contribution. Most importantly, the emotional climate shifts from frustration to pride.

This is atomic productivity in its purest form: small actions, made visible, creating chain reactions of engagement and improvement.

The Willing Worker Assessment

How do you know if you have willing workers trapped in bad systems? After all, people don't inherently show up to work with a mindset of "I don't want to do a good job." Sure, the primary drive for most workers is the paycheck at the end of the week or month, but most want to do a good job. To find your willing workers, look for these signs.

Positive Indicators that willingness exists include high performance when constraints are removed, lots of workarounds (which show initiative), informal collaboration despite formal barriers, ideas shared peer-to-peer even if not upward, and pride in work quality when given autonomy.

Negative Indicators of system suppression include "that's not my job" attitude, minimal effort to meet metrics, high turnover in previously stable roles, cynicism about new initiatives, and knowledge hoarding.

Your Willing Worker Challenge

This week, test the willing worker hypothesis.

Monday: List every barrier your team faces. Tuesday: Ask, "What would you improve if you could?" Wednesday: Remove one barrier (even small). Thursday: Implement one worker suggestion. Friday: Celebrate what happened.

Watch what emerges when willing workers are unleashed.

The Ultimate Truth

Here's what years in various industries have taught me: The vast majority of workers want to do good work; They want to contribute and go home feeling accomplished.

Your job isn't to motivate them. It's to stop demotivating them.

Build systems that trust them, enable them, and respond to them, and you will unlock not just productivity, but passion.

The atomic impact of a willing worker, properly supported, is immeasurable. When you unleash human potential, you don't just get better metrics. You get innovation, engagement, and energy that transforms everything.

The willing workers are already in your organization. The only question is this: will you unleash them?

CHAPTER 14

Leadership for Atomic Impact

Leading systems, not just people

"A system must be managed. It will not manage itself."

- W. Edwards Deming

While this book isn't intended to be another leadership or management book, we can't consider atomic impacts without acknowledging that leadership and management play a critical role in ensuring the systems we design are properly understood, supported, and managed. Systems thinking without leadership is like a perfectly engineered vehicle without a driver that is full of potential but directionless. Leadership, in this context, is less about authority and more about stewardship. This means there is a responsibility to create the conditions for systems to function at their highest potential.

This chapter isn't about traditional leadership traits or executive presence. It's about what kind of leadership behaviors amplify systemic productivity, how managers can influence the feedback loops within systems, and what it means to lead with a systems-oriented mindset.

Systemic Influence vs. Individual Control

One of the most dangerous myths in leadership is that individual performance drives results. While effort and skill absolutely matter, Deming's Red Bead Experiment showed us that even the most competent and willing people are limited by the systems in which they work.

Atomic Impact leadership starts from this premise: don't fix people; fix the system. The job of leadership is not to squeeze more out of people but to remove friction from their work. That's how small wins become atomic catalysts. Considering the goal of *Atomic Impact* is to understand how to get more out of less, or 'squeezing dollars out of dimes." It is critical to understand that the system is where we should squeeze, not the people.

Leading Through Feedback Loops

As discussed in earlier chapters, feedback loops are essential for system regulation. Leaders who fail to recognize this often rely on reactionary decision-making, meaning rewarding short–term gains and punishing anomalies without understanding the underlying structure.

Leaders aligned with systems thinking do the opposite. They listen for patterns before responding to events. They adjust incentives to encourage long-term behavior change. Instead of bonuses for hitting numbers, they create rewards for system improvements that sustain.

Leaders also model reflection by asking, "What is the system telling us?" rather than "Who messed up?" This shift in language changes everything. It moves teams from blame to analysis.

Have you ever heard of the fundamental attribution error? I first learned of this from Dr. Payne from MIT, where he described it as the tendency to attribute outcomes (in the context of system dynamics) to the behavior of individuals rather than the system itself.

For example, if absenteeism rises in a department, the traditional leader might penalize the individuals. A systems leader investigates whether workloads, team dynamics, or unclear policies are the drivers of people missing work. They don't default to blame; they default to analysis.

Trust, Autonomy, and Alignment

One of the most underutilized productivity levers is trust - not blind trust, but engineered trust. This is trust built through systems that reinforce shared goals, visibility, and accountability.

Consider Google's "Project Aristotle," which found psychological safety to be the most important dynamic in high-performing teams. Employees don't need constant supervision; they need environments where it's safe to take initiative and challenge assumptions. This only happens in well-designed systems, and leaders are the ones who make that safety possible.

Here's the critical insight: autonomy is not the absence of oversight; it's the presence of clarity. When people understand how their role connects to broader outcomes, they self-regulate. In this sense, the leader becomes the translator of strategy into system, ensuring that the organizational mission doesn't get lost in silos or ambiguity.

The Atomic Leader's Mindset

To lead for atomic impact, one must abandon the idea of quick wins and instead adopt a mindset of deliberate leverage. Atomic leaders think in systems, act on signals, and build infrastructure for others to thrive. Here are five traits of atomic leadership.

First, **System Awareness.** They map systems, not just org charts. They understand how roles, tools, and processes interact. After learning system dynamics, I have frequently found myself constructing systems dynamics flows to understand the very connection (with feedback loops, stocks and flows, etc.) between everything within a system.

Second, **Curiosity Over Control**. Instead of telling people what to do, they ask how the system is working and what's getting in the way.

Third, **Resilience Through Design.** They design redundancies, buffers, and learning loops into their systems, expecting failure, not fearing it.

Fourth, **Micro-Interventions.** They recognize that small changes, such as changing how a daily huddle works, can yield outsize results.

Fifth, **Long-Term Alignment**. They make decisions that compound, even when they don't produce immediate returns.

Atomic Impact as a Cultural Outcome

Finally, we must understand that the greatest systems in the world won't work if the culture doesn't support them. Leadership is what translates design into behavior and behavior into outcomes. Leaders shape culture through consistency, communication, and how they respond when the system is strained.

Culture is, in many ways, the shadow system, the unwritten code of how things get done. Leaders must align the formal system (policy, process, and KPIs) with the informal system (habits, values, and norms) to avoid productivity breakdowns.

This misalignment is devastatingly common. The formal system might reward speed while the informal culture values quality. Official metrics might track individual performance while actual success depends on collaboration. Policies might encourage innovation while the unwritten rules punish failure. These contradictions create confusion, errors, and cynicism.

When leaders recognize and align both systems, making the formal rewards match the informal values, transformation accelerates. The choice isn't between speed or quality, individual or team success, or innovation or stability. It's about making both systems tell the same story. When they do, performance doesn't just improve; it soars. People no longer waste energy navigating contradictions.

True atomic leadership creates coherence between what organizations say they value and what they actually reward. That alignment is where culture becomes a multiplier rather than a barrier.

Practical Leadership Actions for Atomic Impact

Daily Actions include asking "What prevented excellence today?" not "What went wrong?", removing one barrier (however small), recognizing one system improvement, and connecting one task to a larger purpose.

Weekly Rhythms involve reviewing system metrics, not just outcome metrics, conducting barrier-removal sessions with teams, testing one small system improvement, and sharing learnings across departments.

Monthly Practices include mapping one process with the people who do it, calculating CRC of recent changes, adjusting based on system feedback, and celebrating compound improvements.

Quarterly Imperatives involve auditing systems for accumulated friction, realigning metrics with desired behaviors, investing in system capabilities, not just outputs, and telling stories that reinforce system thinking.

The Leader's Ultimate Question

Here's the question that separates atomic leaders from traditional managers. And I must say, there is a huge difference between the leader and the manager but we won't get into that:

"Am I making it easier for good people to do great work?"

If the answer is yes, you're leading systems, not just people. You're removing friction, not adding oversight. You're multiplying capability, not just managing resources.

In the end, atomic leadership isn't about being the smartest person in the room. It's about creating rooms where everyone gets smarter.

The future belongs to leaders who see systems, cultivate conditions, and unleash the atomic potential that already exists in their organizations.

Will you be one of them?

CHAPTER 15

Creating Psychological Safety

The invisible infrastructure
of atomic productivity

"There's no learning without risk. There's no risk without safety."

This paradox sits at the heart of atomic productivity. For chain reactions to occur, people must try new things. But they'll only try new things if it's safe to fail. This makes psychological safety not just nice to have. It's the invisible infrastructure that enables everything else.

The Safety-Performance Connection

The difference between high-innovation and low-innovation teams often comes down to language. In psychologically safe environments, you hear phrases such as, "Good try, what did we learn?" and

"Interesting idea, how could we test it?" In unsafe environments, the vocabulary is different: "That's not how we do things" and "Whose fault is this?" Same organizations, same resources, completely different psychological climates and completely different outcomes.

Defining Psychological Safety in Atomic Terms

Amy Edmondson defined psychological safety as "a belief that one will not be punished or humiliated for speaking up with ideas, questions, concerns, or mistakes."

In atomic productivity terms, psychological safety is the medium through which chain reactions propagate. Without it, atomic actions fizzle out. With it, they multiply exponentially.

Think of it this way: fear is an insulator that blocks energy transfer. Safety is a conductor that enables energy flow. Trust is the amplifier that multiplies energy.

To understand this deeply, consider what happens at the atomic level. In nuclear physics, chain reactions require three conditions: fissile material (the potential), neutron flow (the trigger), and the right medium (the environment). Remove any element and the reaction stops.

Organizations work the same way. You have willing workers (fissile material) with innovative ideas (neutrons), but without psychological safety (the medium), those ideas never trigger transformation. They're absorbed by fear, deflected by criticism, or simply decay in silence.

When psychological safety exists, something remarkable happens. One person's small improvement (their atomic action) doesn't just succeed, it gives others permission to try. Their success emboldens more attempts. Each innovation lowers the activation energy for the next one. Suddenly, you have a self-sustaining chain reaction of improvement.

The physics parallel goes deeper. In hostile environments, people develop protective shells. Like electrons in stable orbits, they resist change because the energy cost feels too high. In the world of DE&I, I've learned this to be called *covering*. In psychologically safe environments, those same people become reactive, eager to share ideas, combine efforts, and release their potential energy.

That's why psychological safety isn't a "soft," nice-to-have concept. It's the fundamental infrastructure that determines whether your atomic actions create transformation or evaporation. Without it, you're trying to create nuclear power in a vacuum, and as I said in Chapter 1, productivity itself doesn't occur in a vacuum. With it, every small action has the potential to trigger exponential change.

The Chemistry of Trust

Psychological safety doesn't happen by accident. It's created through specific leadership behaviors and system designs. Like a chemical reaction, it requires the right elements in the right proportions.

Catalyst 1: Leader Vulnerability. When leaders admit mistakes, ask for help, and say "I don't know," these signal that imperfection is acceptable. Leadership vulnerability isn't a sign of weakness, but a sign to truth and reality.

Catalyst 2: Process Clarity. Ambiguity breeds fear. Clear processes and expectations create safety to operate within boundaries.

Catalyst 3: Rapid Response. How quickly leaders respond to safety violations determines future safety levels.

Catalyst 4: Celebration Structure. What gets celebrated gets repeated. Celebrating learning, not just success, creates safety to experiment.

Now, it is important to note that these catalysts are important guides to create psychological safety. However, it is also to consider that the leader must be genuine in their efforts and intent. There are many books and publications that tell us how to be effective leaders, talking about what to do, what not to do, how to stand, how to speak, and even how to dress for presence. All of that is relevant and good, but disingenuous actions are easily seen my those we lead.

Building Safety Infrastructure

Creating psychological safety requires intentional infrastructure. Much like a nuclear reactor, the infrastructure to safely let the energy fly is just as important as the actions taking place within the reactor.

Physical Infrastructure includes spaces for private conversations, visual displays of values and norms, comfortable areas for informal interaction, and removal of surveillance that feels punitive. One company removed cameras from break rooms after realizing they prevented honest conversations about improvements.

Process Infrastructure encompasses anonymous suggestion systems, structured experimentation time, clear escalation paths, and protected discussion forums.

Social Infrastructure involves peer support systems, mentorship programs, cross-functional relationships, and celebration rituals. We must define these and ensure they stay intact with true intent.

Digital Infrastructure includes platforms for idea sharing, transparent project tracking, learning repositories, and success story distribution.

These four infrastructures work like the interlocking systems of a reactor. Remove any one and the entire system weakens. Physical spaces without supportive processes create empty gestures. Digital platforms without social trust become unused tools. Process improvements without cultural celebration fade quickly.

The magic happens when all four infrastructures reinforce each other. The anonymous suggestion box (process) leads to ideas shared on digital platforms (digital), discussed in comfortable break rooms (physical), and celebrated at team gatherings (social). Each infrastructure element amplifies the others, creating an environment where psychological safety isn't just possible but inevitable.

Most organizations invest heavily in one type of infrastructure while ignoring others. They build beautiful collaboration spaces but maintain punitive processes. They create digital platforms but forget human connection. They design careful processes but neglect the physical environment. This imbalance explains why so many psychological safety initiatives fail to create lasting change.

The organizations that achieve true psychological safety understand this: infrastructure isn't about any single element being perfect. It's about all elements working together to send one consistent message: your ideas, questions, concerns, and learning are not just welcomed but essential. When that message comes through every channel consistently, psychological safety transforms from a concept to a lived reality.

Overcoming Safety Blockers

Several factors consistently destroy psychological safety. Much like organizations investing in single safety infrastructures, the blockers that exist stifle what happens. Overcoming these blockers is critical to ensure that psychological safety lives and thrives in your organization.

Blocker 1: Blame Culture. When mistakes lead to punishment, people hide problems until they explode. The antidote is to separate system failures from personal failures. Ask, "What about our system allowed this?" not, "Who messed up?"

Blocker 2: Forced Ranking. When employees are ranked against each other, collaboration dies. While competition thrives, it is only celebrated at the top of the ranking. The antidote is to measure team outcomes and system improvements, not individual competition.

Blocker 3: Shooting Messengers. When bearing bad news brings retribution, critical information goes underground. The antidote is to publicly thank people who surface problems early. Make it heroic, not career-limiting.

Blocker 4: Perfectionism. When only perfect outcomes are acceptable, innovation stops. The antidote is to celebrate rapid learning over flawless execution. Perfect is the enemy of better.

Creating Safety Chain Reactions

The most powerful aspect of psychological safety is its multiplicative effect. Since one of the goals for *Atomic Impact* is its multiplicative effect, so too is the benefit of psychological safety.

Individual Impact occurs when one person feels safe to suggest improvement. Team Impact follows as others see the positive response and share their ideas. Department Impact spreads as success stories travel and more teams experiment. Organizational Impact emerges as innovation becomes the norm, not the exception. System Impact embeds continuous improvement in culture.

This progression isn't linear or predictable. It's exponential and self-reinforcing. Each level of impact reduces the activation energy for the next level. When individuals see their ideas implemented, teams become bolder. When teams succeed, departments take notice. When departments transform, the organization shifts. When the organization demonstrates sustained change, it becomes embedded in the cultural DNA.

The multiplication happens because psychological safety removes the friction that typically slows or stops the spread of innovation. In unsafe environments, good ideas die in isolation. In safe environments, they become catalysts for chain reactions that transform everything they touch.

The Leader's Safety Checklist

Leaders serious about creating psychological safety should regularly assess their daily behaviors. Do I respond to mistakes with curiosity or criticism? Do I ask questions that invite truth or compliance? Do I model vulnerability or perfection? Do I celebrate learning or just results?

Weekly practices matter, too. Have I publicly admitted a mistake and shared learning? Have I thanked someone for disagreeing with me? Have I protected someone who took a smart risk that failed? Have I asked, "What are we not talking about that we should?"

Monthly systems require attention as well. Are our metrics driving fear or innovation? Are our processes enabling or constraining? Are our communications building or breaking trust? Are our celebrations reinforcing safety or competition?

Your Safety Building Challenge

This week, build psychological safety through one specific action each day.

Monday: Share a personal failure and what you learned from it. Tuesday: Ask, "What would you try if you knew you couldn't fail?" Wednesday: Thank someone publicly for surfacing a problem. Thursday: Change one process that punishes honest mistakes. Friday: Celebrate a smart failure as enthusiastically as a success.

The Compound Effect of Safety

Here's what's remarkable about psychological safety: it compounds faster than any other investment in organizational capability.

When people feel safe, they share ideas freely, which drives innovation multiplication. They admit mistakes quickly, catching problems early. They help others openly, spreading knowledge rapidly. They challenge assumptions, improving systems continuously. They bring full selves to work, increasing energy exponentially.

Each safe interaction makes the next one more likely. Each innovation attempted makes the next one easier. Each problem surfaced early prevents ten later.

That's atomic productivity through psychological safety - not just enabling chain reactions, but creating an environment where they happen naturally, continuously, and exponentially.

The Ultimate Safety Test

Want to know if you've created real psychological safety? Look for these signs:

Bad news travels faster than good news. Experiments outnumber perfect executions. Questions outnumber assertions in meetings. People challenge ideas regardless of hierarchy. Failures are dissected publicly for learning. Innovation comes from unexpected sources.

But here's the ultimate test: can your newest, most junior employee challenge your oldest, most senior leader's idea in a meeting?

If yes, you've created psychological safety. If no, you're leaving atomic potential untapped.

Psychological safety isn't about being nice. It's about being effective. It's the difference between a system that learns and adapts and one that stagnates and decays.

In atomic productivity, psychological safety isn't overhead; it's the operating system. Without it, you're just managing tasks. With it, you're unleashing transformation.

The choice is yours.

PART VI:

Real-World Applications

CHAPTER 16

Atomic Productivity in Crisis

How extreme pressure reveals the power of small actions

"It was the best of times, it was the worst of times." Dickens could have been describing the COVID-19 pandemic's impact on productivity. While some organizations crumbled, others discovered atomic actions that transformed their operations forever.

Crisis strips away the non-essential. It reveals which systems are robust and which are facades. Most importantly for our purposes, crisis proves that atomic productivity isn't just a nice-to-have optimization strategy, it's a survival mechanism.

The Pandemic Laboratory

COVID-19 created a natural experiment in organizational adaptability. Within the same industries, facing identical challenges, some organizations thrived while others barely survived. The difference was not in resources or market position but in response patterns at the atomic level.

Traditional crisis response creates rigid command structures, compliance-focused protocols, and punitive measures for deviation. The result is predictable: increased stress, mass departures, customer dissatisfaction, and lost market share.

Atomic crisis response distributes decision-making, provides frameworks for local adaptation, focuses on outcomes over process, and celebrates innovation. These organizations see resilient operations, retained staff, customer loyalty, and gained market share.

Same crisis, opposite outcomes. The difference is in the details, the atomic actions.

The Crisis Amplification Effect

Here's what's counterintuitive: crisis doesn't diminish the power of atomic actions. It amplifies them. When resources are scarce and pressure is high, small improvements create disproportionate impact.

Why? Three reasons emerge.

First, Reduced Margin for Error. When you're operating at capacity, tiny improvements prevent system collapse. A five-minute efficiency gain during normal times is nice. During crisis, it's the difference between serving all customers and having riots in your parking lot. (Yes, I witnessed this.)

Second, Heightened Awareness. Crisis makes people pay attention. Atomic actions that might be ignored in calm times get noticed, adopted, and spread rapidly. In my dissertation research and through experience, business got back to the "nuts and bolts," finding and focusing on what truly mattered.

Third, Survival Motivation. When survival is at stake, resistance to change evaporates. People try things they'd never consider normally. Take the example of my role as the Recruiting and Training Manager at Kroger. In an instance to ensure we properly supported customers and team members falling ill with the virus, we hired 3,000 temporary workers, something that previously hadn't been considered.

Atomic Actions That Emerged During COVID

Crisis revealed atomic actions hiding in plain sight. Simple floor markings could eliminate crowding and conflict with minimal investment. Reserved shopping hours for vulnerable populations created safety while building community goodwill. Pairing employees for mutual support reduced callouts and improved morale exponentially. Brief gratitude acknowledgments at shift starts transformed energy and effort in ways metrics couldn't capture.

These weren't complex strategies. They were atomic actions with crisis-amplified CRCs estimated at 25:1. The same actions suggested during normal times might have been ignored. Under pressure, their value became undeniable, and adoption spread at viral speeds.

The Three Phases of Crisis Productivity

Through observing multiple organizations navigate COVID, I identified three distinct phases, each requiring different atomic strategies.

Phase 1: Shock (Weeks 1–4) was characterized by panic, confusion, and reactive decisions. The atomic focus was on stabilization actions. Key moves included simplifying everything, communicating constantly, and preserving energy. The success metric was survival. Example atomic actions included five-minute stand-ups for critical updates only, single-point decision makers for rapid response, suspending all nonessential processes, and constant communication for changing protocols.

Phase 2: Adaptation (Months 2–6) was characterized by learning, experimentation, and new patterns. The atomic focus shifted to innovation actions. Key moves involved testing rapidly, failing fast, and scaling what works. The success metric was resilience.

Phase 3: Evolution (Months 6+) was characterized by new normal, sustainable systems, and growth. The atomic focus moved to optimization actions. Key moves included embedding learnings, building on strengths, and preparing for the next crisis or strain of COVID. The success metric was transformation.

The Psychology of Crisis Productivity

Crisis fundamentally alters human psychology in ways that can either destroy or amplify productivity.

The Destruction Path follows Fear → Freeze → Minimal compliance → System breakdown.

The Amplification Path follows Purpose → Focus → Innovation → System breakthrough.

The atomic actions that work in crisis all nudge people toward the amplification path. They connect every action to serving others through purpose. They give control where possible through autonomy. They enable rapid skill development through mastery. They make improvements visible through progress. They foster mutual support through community.

Building Crisis-Ready Atomic Systems

You can't wait for a crisis to develop atomic capabilities or even point you in the direction of simple improvement. This reality has spawned an entire industry of business continuity planning, with certifications like CBCP (Certified Business Continuity Professional) and frameworks like ISO 22301. These programs exist because organizations learned the hard way that crisis preparation can't be improvised. Had it not been for an opening at the college I teach at, I would not have known such a field existed

However, traditional business continuity often focuses on maintaining operations exactly as they were before disruption. This approach misses the transformative potential of crisis. Atomic crisis preparation is different because it builds adaptability, not just continuity.

Pre-crisis preparation determines crisis performance. Organizations must first identify their essential atomic actions, which represent the 20 percent of activities that drive 80 percent of their value. Once identified, these critical actions need supporting infrastructure that creates flexibility rather than rigidity. This includes comprehensive cross-training matrices that ensure knowledge doesn't reside in single individuals, clear decision delegation protocols that function even when leadership is unavailable, and strategic resource buffers that provide breathing room during disruption.

Beyond infrastructure, organizations need to practice adaptation through controlled chaos. Regular crisis simulations and innovation challenges would build the needed organizational muscle memory needed to respond creatively under pressure. These exercises reveal hidden dependencies and strengthen the sensing systems that provide early warning of disruption. When front-line feedback channels are robust and trusted, organizations can detect and respond to changes before they cascade into crises.

The prepared organization doesn't try to predict specific crises, which is impossible. Instead, it builds the capability to handle any disruption through atomic adaptability. This approach transforms a crisis from a threat to be endured into an opportunity for discovering better ways of operating.

Your Crisis Preparation Audit

Assess your organization's crisis readiness through an atomic lens.

System Flexibility: Can your core processes function at 50 percent staffing? Can decisions be made if key leaders are unavailable? Can you pivot operations within forty-eight hours?

Human Resilience: Do people understand the "why" behind their work? Are there support systems for personal challenges? Is there psychological safety to voice concerns?

Innovation Capacity: Can front-line workers test improvements? Are there mechanisms to spread good ideas rapidly? Do you celebrate learning from failure?

Communication Robustness: Can critical information reach everyone within hours? Are there feedback loops from field to leadership? Do people trust the information they receive?

The Silver Lining of Crisis

Here's what's remarkable: many organizations never want to go through another pandemic, but they also never want to lose what they learned. Crisis forced them to discover atomic actions they should have found years earlier.

Common permanent changes from COVID include simplified processes that work better, distributed decision-making that responds faster, human connections that run deeper, innovation muscles that stay strong, and resilience that handles any challenge.

The tragedy isn't that it took a pandemic to force these discoveries. It's that most organizations will slowly drift back to their old ways once memory fades. The companies that thrive post-crisis are those that institutionalize their emergency innovations before bureaucracy creeps back in. They recognize that the simplified processes weren't just crisis workarounds; they were better ways of working all along. The distributed decision-making wasn't just pandemic necessity; it was always faster and more effective than centralized control. These organizations treat their crisis learnings as a competitive advantage to protect, not an emergency measure to abandon.

The Next Crisis Is Coming

We don't know what the next crisis will be. Economic collapse? Climate disaster? Technology disruption? Geopolitical upheaval? The specific crisis matters less than your atomic readiness for it.

Organizations with atomic productivity capabilities don't just survive crises, they use them as catalysts for transformation. They emerge stronger, more innovative, and more human.

The question isn't whether another crisis will come. It's whether you'll be ready to find the atomic actions within it that transform challenge into opportunity.

Start building that capability today because in crisis, there's no time to learn atomic productivity. There's only time to apply it.

And those who can, will inherit the future.

CHAPTER 17

Case Studies from the Frontlines

Stories from the simulation lab

Numbers tell what happened. Stories tell why it mattered. In this chapter, I'll share detailed case studies that started as late-night experiments in computational modeling and turned into revelations about organizational transformation.

You know how I mentioned my fascination with *Modern Marvels* and *How It's Made*? Well, that same curiosity that had me watching factory lines and engineering marvels led me down a rabbit hole of organizational modeling. After diving deep into quantum theory and atomic energy, I couldn't shake this question: could we model organizational change the same way physicists model particle interactions?

So, I taught myself computational modeling and started building simulations. Think of these like organizational video games where you can test "what if" scenarios without risking real jobs or companies. Each case study here started as agent-based models and Monte Carlo simulations running thousands of scenarios to see what really drives transformation.

The beautiful thing about simulations is that they strip away the noise and show you the patterns. Like watching those Modern Marvels episodes where they speed up construction footage, suddenly you see the design principles that matter.

Why Simulations Matter for Atomic Productivity

Traditional business books rely on anecdotes and best practices from successful companies. Often successes are hyper-focused while other parts are not as successful, but correlation isn't causation. Just because Company X succeeded while doing Y doesn't mean Y caused the success. Simulations let us test causation by running thousands of scenarios with controlled variables.

These models revealed the mathematical foundation of atomic productivity. They proved that small actions in the right places create predictable, exponential improvements. They showed which environmental factors amplify atomic actions and which suppress them. Most importantly, they demonstrated that transformation follows consistent patterns regardless of industry or context.

Case Study 1: The Midnight Miracle

When the simulation predicted the impossible

I built this first simulation to test whether "bad employees" were really bad or just trapped in bad systems. The model included 150 digital workers, each with attributes such as error rates, turnover probability, and productivity scores. Then I started tweaking environmental variables: lighting quality, break room comfort, recognition frequency, and manager presence on the floor.

The initial state matched common midnight shift realities: 89 percent accuracy (when 97 percent was standard), 200 percent annual turnover, and productivity at 60 percent of day shift.

After running 10,000 simulations with different intervention sequences, the results were shocking. The same digital agents achieved 98.5 percent accuracy when environmental factors improved. Not new people. Same agents, different parameters.

What the Model Revealed

The simulation uncovered a simple but powerful relationship: error rates doubled in poor lighting conditions. When lighting improved from 30 percent to 80 percent adequacy, errors dropped 15 percent immediately. This wasn't about motivation or skill. It was physics. People can't pick what they can't see.

Environment was just the start. The model showed that manager presence on the floor created network effects. Information spread faster, trust metrics improved, and innovation adoption accelerated exponentially. The mathematical relationship was clear: trust grew at a compound rate based on manager floor time.

Small interventions cascaded through the system. Free coffee and snacks improved satisfaction scores. Improved satisfaction reduced turnover probability exponentially. Lower turnover meant retained knowledge. Retained knowledge improved accuracy. Each atomic action triggered others.

After 90 days in the simulation, median outcomes across thousands of runs showed: accuracy improving from 89 percent to 98.5 percent, turnover dropping from 200 percent to 15 percent annually, productivity increasing 40 percent, all from investments totaling less than $1,000. The ROI exceeded 4,500 percent.

The simulation taught me something models can't fully capture: we weren't modeling workers. We were modeling people who wanted to belong to something that mattered.

Case Study 2: The Call Center Phoenix

System dynamics meets human dignity

This simulation explored whether call centers were doomed to high turnover and low satisfaction, or if the system itself was the problem. I built a system dynamics model with interconnected feedback loops. Picture a bathtub with multiple faucets and drains: experience flowing in through training, draining out through turnover; satisfaction filling from good interactions, emptying from bad systems; innovation accumulating from suggestions, evaporating from ignored ideas.

The initial conditions modeled a typical failing center: 11 systems to navigate, 42 percent customer satisfaction, 180 percent annual turnover, and eighteen-minute average handle time.

Then I introduced changes, fFirst, suspending all metrics except satisfaction. In the model, this was simply changing stress calculations from compound penalties to reduced pressure. The simulation went wild. Freed from handle-time pressure, agents started actually solving problems. First-call resolution jumped. Repeat calls dropped. Total call volume decreased.

The Cascade Effect

The simulation revealed fascinating dynamics. When I modeled peer mentoring as experience transfer between agents, the knowledge distribution curve shifted dramatically. New agents learned faster. Experienced agents felt valued. System-wide capability improved exponentially.

Consolidating from eleven systems to one had profound effects. The mathematical relationship was stark: each additional system added seven seconds to handle time. Eleven systems meant seventy extra seconds per call. Multiplied by thousands of calls, the waste was staggering.

A 90-day simulation results showed customer satisfaction rising from 42 percent to 67 percent, first-call resolution improving from 31 percent to 58 percent, voluntary overtime jumping from 0 percent to 40 percent, handle time decreasing 30 percent, and annual savings of $4.2 million from reduced turnover and improved efficiency.

The simulation confirmed what we know intuitively: people trust peers more than procedures. When agents created training content for each other, knowledge retention tripled compared to top-down training.

Case Study 3: The Retail Renaissance

When community connection beats corporate policy

This simulation started personal. My wife's family is from northern New Mexico, a small town where everyone knows everyone. I wondered whether community connection was actually measurable. What if social bonds affect business metrics?

I built a spatial agent-based model simulating fifty employees and 500 daily customers, all with connection networks. As a twist I modeled social pressure. If you know someone who works at the store, your theft probability changes based on the strength of that connection.

The initial state modeled a failing store: 8 percent shrink rate, 300 percent turnover, and declining sales. Then I simulated hiring locally. Local employees had three times more community connections. Each connection reduced theft probability. The network effects were exponential.

The Dignity Economics

The simulation revealed something profound about human dignity. When I modeled dignity as a performance multiplier, small improvements created compound effects. Better break rooms, employee music choice, and recognition walls each increased dignity scores that multiplied all other performance metrics.

The "Local Heroes" wall was just a variable in the model, but it triggered remarkable dynamics. Customers with pictures on the wall had 90 percent lower theft rates. Their networks showed five times more positive interactions. Pride proved contagious in measurable ways.

Here a the six-month simulation outcomes: shrink dropping from 8 percent to 1.5 percent, sales increasing 120 percent, turnover falling from 300 percent to 25 percent, annual profit increase of $2.8 million, all from investments under $25,000.

The spatial model revealed spillover effects. As the store improved, crime in surrounding blocks dropped 30 percent in the simulation. Other businesses opened nearby. Property values increased. The store became a positive feedback loop for the entire neighborhood.

Case Study 4: The Hospital Transformation

Discrete events, continuous impact

Healthcare fascinated me because errors can cost lives. In 2023 I lost my mother. While not specifically due to an error, many of her symptoms were caused by errors. I built a discrete event simulation modeling patient flow through hospital systems with 200 nurses accumulating fatigue, 1,000 patients per week with varying acuity, error probability calculated as a function of fatigue and system complexity, and turnover events following exponential distributions.

The initial state showed 45 percent annual turnover, bottom 10 percent patient satisfaction, rising medical errors, and seven miles walked per shift per nurse.

I modeled various interventions: supply caches at point-of-care, voice-to-text documentation, protected patient time, and gratitude rounds. Each had specific effects on the system dynamics.

Removing Friction, Adding Life

The simulation quantified friction removal precisely. Supply caches eliminated 40 percent of walking. Less walking meant more energy. More energy meant better attention. Better attention prevented errors. The mathematical relationship was clear and compelling.

Voice-to-text documentation cut administrative time 50 percent. That time became patient time. Patient time improved satisfaction. Satisfaction reduced turnover. Everything connected in measurable ways.

Sacred hour (uninterrupted patient time) showed 30 percent error reduction during those periods. The simulation revealed why: task-switching has a cognitive cost that accumulates. Interruptions don't just pause care, they degrade it.

Here are the six-month results: turnover dropping from 45 percent to 8 percent, patient satisfaction jumping from 10th to 75th percentile, errors reduced by 60 percent, estimated lives saved: 47, investment under $100,000, annual savings exceeding $4 million.

The Patterns in the Code

After running hundreds of these simulations, undeniable patterns emerged:

Leadership Proximity Matters: Every simulation showed the same mathematical relationship: transformation speed was inversely proportional to hierarchical distance. The closer leaders were to front-line work, the faster positive changes propagated.

Dignity Has ROI: Across every model, dignity investments showed exponential returns. Treating people as humans didn't just feel right; it multiplied every other metric.

Networks Beat Hierarchies: Innovation spread followed network topology, not org charts. Peer connections predicted adoption speed better than any other variable.

Small Actions, Big Effects: Every simulation showed power law distributions. 20 percent of actions drove 80 percent of results, but identifying the right 20 percent required modeling.

Patience Pays Exponentially: Short-term costs led to long-term exponential gains. Every model showed the same curve: initial investment, flat period, then explosive growth.

The Truth About Transformation

These simulations taught me something profound: transformation isn't magic; it's math. Small actions compound predictably. Systems thinking beats people blaming consistently. Environment shapes behavior reliably.

Here's what models can't capture: the moment when a midnight shift worker wears their pride to their kid's school, when innovation saves a life, when a community rallies around a store that rallied around them.

The simulations show it's possible. The math proves it's profitable. The stories remind us why it matters.

Your organization has atomic potential waiting to be unleashed. These simulations prove that small, strategic actions in the right places create exponential transformation. The patterns are consistent. The math is compelling. The only question is whether you'll apply these principles to unlock your organization's atomic potential.

CHAPTER 18

The Future
of Work

How atomic productivity will
reshape everything

"The future is already here – it's just not evenly distributed."

- William Gibson

As I write this in 2025, we're standing at an inflection point. The convergence of AI, automation, demographic shifts, and the lessons from COVID-19 and The Great Resignation have created a perfect storm for workplace transformation. Here's what most futurists miss: the future of work isn't about the big technological leaps. It's about the atomic actions that will determine whether those leaps help or harm humanity.

The Three Futures Before Us

Based on my research and observation of emerging trends, I see three possible futures.

Future 1: The Dystopian Path sees AI and automation eliminate jobs faster than new ones emerge. Wealth concentrates in the hands of technology owners. Workers become gig-economy servants to algorithms. Human creativity and connection are monetized and extracted. Productivity gains benefit only a few.

Future 2: The Utopian Dream envisions technology liberating humans from drudgery. Universal basic income supports creative pursuits. Work becomes optional and meaningful. Human potential is fully unleashed. Abundance is shared equitably.

Future 3: The Atomic Reality recognizes technology amplifies human capability. Work evolves to focus on uniquely human contributions. Small actions create exponential value. Systems are designed for human flourishing. Productivity gains are shared through smart design.

I believe Future 3 is not only possible but probable if we make the right atomic choices now.

The Emerging Atomic Workplace

Several trends are converging to create unprecedented opportunities for atomic productivity. It is up to us to acknowledge these trends and align them where we see fit... with intent.

Trend 1: Distributed Intelligence AI isn't replacing human intelligence; it's augmenting it. The winners will be organizations that create human-AI partnerships where each amplifies the other's strengths. When humans maintain decision authority while AI processes vast variables, productivity gains will likely exceed 60 percent with reduced strain on workers.

Trend 2: Micro-Learning Revolution The half-life of skills is shrinking. Traditional education can't keep pace. But atomic learning delivered in five-minute daily bursts, immediately applied and shared with peers, can show twenty times higher ROI than traditional training programs.

Trend 3: Purpose-Driven Productivity The Great Resignation taught us that people won't tolerate meaningless work. Organizations connecting every atomic action to visible impact on real people should see 30 percent productivity increases, but more importantly, meaning returns to work.

Trend 4: Network Effects of Knowledge In the atomic workplace, knowledge doesn't flow down hierarchies; it spreads through networks. When organizations enable any employee to share innovations that make work easier, they typically implement thousands of micro-improvements annually, each building on others.

These four trends aren't developing in isolation. They're converging to create entirely new ways of working that don't fit traditional job descriptions. As organizations discover the power of distributed intelligence, micro-learning, purpose-driven productivity, and networked knowledge, they're realizing they need people who can orchestrate these forces.

The old roles of manager, analyst, and coordinator assume a world of linear processes and hierarchical control, but atomic productivity requires something different. It requires people who can see patterns others miss, remove friction others accept, design chain reactions others can't imagine, and nurture the human energy that makes it all possible.

These needs are giving birth to roles that didn't exist five years ago and may be essential five years from now.

The New Atomic Roles

Traditional job titles are becoming obsolete. New roles are emerging that embody atomic principles, focusing on multiplication rather than management, systems rather than silos. While we have not formally defined these roles, find them in your organization and watch atomic productivity happen.

The Pattern Spotter lives at the intersection of data and intuition. They identify atomic actions with high CRC potential that others miss, connect dots across departments to reveal hidden leverage points, and surface system dynamics that create unexpected bottlenecks or opportunities. These individuals often come from unconventional backgrounds because pattern recognition transcends traditional expertise.

The Friction Hunter operates like an organizational detective. They systematically identify and remove barriers to productivity, measure the true cost of time and energy waste, and design elegant solutions that feel obvious in retrospect. Their superpower is seeing complexity that everyone else has normalized and knowing how to eliminate it.

The Chain Reaction Designer thinks in systems and sequences. They architect workflows that multiply human effort rather than just organizing it, calculate and optimize CRCs across entire value chains, and build feedback loops that make improvement self-sustaining. These roles often emerge from engineering or operations backgrounds but require a uniquely human understanding of motivation.

The Energy Catalyst serves as the organizational physicist of human potential. They maintain and monitor team psychological safety like a vital sign, facilitate innovation sessions that actually produce innovations, and ensure atomic wins get celebrated and replicated. Without them, even the best atomic systems lose momentum.

These roles rarely appear on job boards because organizations don't yet know they need them. But once one person starts operating in these ways, demonstrating the multiplication possible, the roles spread organically. They represent the future of work: not managing resources but unleashing potential.

The Gig Economy Evolution

The gig economy started as a promise of freedom and flexibility. What we're witnessing now is a fundamental divergence that will determine whether millions of workers thrive or merely survive. The platforms making choices today are creating either poverty traps or prosperity engines. The gig economy is splitting into two paths.

Path 1: Algorithmic Exploitation treats workers as interchangeable units. Algorithms optimize for extraction. It becomes a race to the bottom on wages with no learning or growth.

Path 2: Atomic Entrepreneurship (keeping just the positive path) Workers become micro-business owners, platforms amplify individual capabilities, and value-based pricing emerges with continuous skill evolution. Top performers on platforms that connect specialized expertise with business needs often earn ten times the average wages while delivering fifty times the value through atomic expertise.

The organizations choosing Path 2 are seeing extraordinary results. One platform connecting specialized experts with businesses found that their top performers earned ten times the average wages while delivering fifty times the value through atomic expertise.

Preparing for the Atomic Future

The trends are clear. The new roles are emerging. Knowing where we're headed isn't the same as being ready for it. Most people feel overwhelmed by the pace of change, unsure whether their skills will remain relevant next year, let alone next decade. That anxiety is valid, but it misses a crucial truth: the atomic future isn't about predicting which technologies will dominate or whether our teams will lose their jobs to technology. It's about developing capabilities that multiply value regardless of the tools.

For Individuals: Building Your Atomic Capability

The path forward doesn't require becoming a technologist or abandoning your expertise. It requires developing what I call Atomic Awareness, the ability to see multiplication potential in everyday work. Start tracking which of your activities create ripple effects versus those that die in isolation. You'll quickly discover that 80 percent of your time goes to low-CRC busy work. That's not judgment; that's opportunity.

Building Learning Velocity becomes non-negotiable when knowledge has an expiration date, but this isn't about consuming more content. Master the art of micro-learning: five minutes of focused practice beats hours of passive consumption. More importantly, teach what you learn immediately. Nothing accelerates learning like having to explain it to someone else.

The human superpowers: emotional intelligence, creative problem-solving, systems thinking, and relationship building aren't soft skills anymore. They're the hard differentiators that no algorithm can replicate. (AI won't become sentient.) While others chase technical certifications, invest in capabilities that become more valuable as automation advances.

Here's the reality check: none of this matters if you're trapped in a role that suppresses atomic potential. Start small. Negotiate for autonomy over one process. Propose one system improvement. Document your multiplication effects. Build evidence that your atomic actions create exponential value. Then use that evidence to design your next role. But consider this. Focus and master your current role. If you follow these atomic practices, the next role will come.

For Organizations: The Transformation Imperative

Organizations face a starker choice than individuals. The comfortable fiction that incremental improvement will suffice is dead. Atomic transformation isn't optional; it's existential.

Redesigning for Multiplication starts with brutal honesty. Audit every process through the CRC lens. You'll find that most organizational activity is expensive theater; meetings about meetings, reports nobody reads, and approvals that add no value. Eliminating friction isn't about working faster; it's about stopping work that shouldn't exist.

Investing in Human Amplification requires abandoning the automation-as-replacement mindset. The highest returns come from AI that makes humans superhuman, not unemployed. This means tools that eliminate tediousness while preserving judgment, platforms that spread innovations naturally, and systems that capture and multiply every improvement.

Building Adaptive Capacity challenges every assumption about planning and control. Regular experimentation rhythms matter more than perfect strategies. Fast failure loops teach more than careful analysis. Distributed decision authority responds faster than hierarchical approval. Resilience beats efficiency when the only constant is change.

This is where most organizations fail. None of this works without "Sharing the Gains." If atomic contributors watch all value flow to shareholders while their wages stagnate, the system breaks. You must link rewards directly to value multiplication. Create equity participation that makes atomic success everyone's success. Celebrate system improvements as enthusiastically as sales wins.

The Global Implications

Let's be honest: these aren't just nice ideas for progressive companies. The implications are civilizational. I mean, many organizations do have global reach right?

Consider the Economic Impact. If every organization improved productivity just 20 percent through atomic actions, entirely achievable based on our case studies and simulations, global GDP could increase by trillions without consuming additional resources. I once saw this happen when working at Kroger. We implemented a process that focused on what would classify as atomic actions, capturing millions of dollars without additional resources. This isn't growth through exploitation but through optimization. It's the only sustainable path forward.

The environmental benefit is equally profound. Atomic productivity means doing more with less, which is exactly what our planet demands. Organizations implementing these principles can consistently report 30–50 percent waste reduction (anecdotally of course), not through green initiatives but through system optimization that happens to be green.

Social Equity emerges naturally when productivity comes from system design rather than individual heroics. When success depends on atomic actions rather than pedigree, opportunities expand for everyone. The atomic workplace doesn't care about your background; it cares about your contribution to the chain reaction. We measure contribution when it comes to finance. Why not ensure contribution is measured at the atomic level?

This leads us to genuine Human Flourishing. When work leverages creativity and eliminates drudgery, mental health improves. When people see their atomic actions create real impact, purpose returns. When communities benefit from productive organizations, social fabric strengthens.

The Choice Before Us

We need to be clear-eyed about the alternatives. This isn't academic theory (well some is); it's an urgent choice with radically different outcomes.

The Old Path leads somewhere we can already see: algorithmic management squeezing every drop of productivity from human cogs, wealth concentrating in ever-fewer hands while workers compete for scraps, technology surveilling and controlling rather than enabling, and productivity gains that come at the cost of human dignity. This path ends in social breakdown.

The Atomic Path offers something different but requires courage to choose: systems designed for human flourishing, not just efficiency; technology that amplifies human capability rather than replacing it; prosperity shared through design, not trickled down through charity; and productivity that comes from multiplication, not exploitation.

Your Future Starts Now

Here's the uncomfortable truth: the atomic future won't wait for you to be ready. It's already emerging in thousands of organizations worldwide. Every day you delay is a day your competitors advance.

Here's the empowering truth: you don't need anyone's permission to begin. The next time you simplify a process, you're building the atomic future. When you share knowledge that multiplies someone else's capability, you're creating chain reactions. When you design out friction or celebrate learning or question a wasteful practice, you're not just improving your workplace. You're proving that the atomic path works.

The Atomic Manifesto for Future Work

This isn't wishful thinking. It's an observable reality. Small actions, designed well, do change everything; we've proven it mathematically and empirically. Systems do determine outcomes more than individual effort; every simulation confirms it. Technology should amplify humanity, not replace it. The highest CRCs always involve human-machine partnership.

Knowledge shared multiplies exponentially. We can see it in every networked organization. Purpose and productivity aren't opposites but partners; meaningful work consistently outperforms mechanical work. The future is human plus machine, not human versus machine. The evidence is overwhelming.

Every person has atomic potential waiting to be unleashed. Organizations that multiply human capability will inherit the future, they're already outcompeting those that don't. Sustainability comes from doing more with less, better, it's the only path that scales.

The best time to start was yesterday. The next best time is now.

The Call to Action

The future of work won't be determined by Silicon Valley visionaries or Wall Street calculations. It will be determined by millions of atomic actions taken by people exactly like you, in unremarkable workplaces, on ordinary Tuesday afternoons.

You don't need an innovation lab. You don't need R Studio. You don't need a transformation budget. You don't need executive approval. You just need to see the atomic potential that already surrounds you.

While others debate the future of work, you can be creating it, one atomic action at a time starting with your very next decision.

The future of work isn't a prediction. It's a choice, and that choice is yours.

What will your first atomic action be?

PART VII

Advanced Concepts

CHAPTER 19

Your First 100 Days of Atomic Transformation

From theory to reality:
A practical implementation guide

"Everyone has a plan until they get punched in the mouth."
- Mike Tyson

You've read the theory. You understand the principles. You're inspired by the possibilities. Now comes the hard part; making it real in your organization. The gap between understanding atomic productivity and implementing it is where most transformations die, not because the concepts don't work, but because the path from here to there seems impossibly complex. It doesn't have to be.

Over the past several years, I've been part of and watched organizations of every size and type attempt atomic transformation. The ones that succeed follow remarkably similar patterns. They start small, build momentum systematically, and most importantly, they follow a sequence that makes change feel inevitable rather than impossible. Here is the paradox. They just didn't know they were operating and transforming in an atomic capacity.

This chapter is your implementation roadmap. Instead of abstract principles it provides concrete steps you can take starting tomorrow morning. Atomic transformation isn't about the perfect plan. It's about the first action that starts the chain reaction.

Why Most Transformations Fail

Before we dive into what to do, let's be honest about what you're up against. Most organizational transformations fail. The statistics are brutal: 70 percent of change initiatives don't achieve their goals. Here's what those statistics don't tell you: they're measuring the wrong kind of change.

Traditional transformations fail because they try to change everything at once. They launch with fanfare, demand immediate compliance, measure lagging indicators, and exhaust everyone involved. Let's not forget about the KPIs tied to all of this. By month three, people are cynically waiting for the initiative to pass like a kidney stone.

Atomic transformation is different. It starts invisibly, builds through attraction, not mandate, measures leading indicators of multiplication, and energizes participants through visible wins. Most importantly, it doesn't feel like change. It feels like improvement.

The secret is in the sequence. Just as nuclear chain reactions require precise steps to achieve criticality, organizational transformation requires the right actions in the right order. Get the sequence wrong and nothing happens. Get it right and transformation becomes inevitable.

Days 1–10: Assessment and Quick Wins

Your first ten days aren't about changing anything. They're about seeing clearly. Most organizations are so busy fighting fires they've never mapped where the heat comes from. This is your chance.

Start with shadow observation. Pick three different roles in your organization, and shadow the people doing them for two hours each. Don't interview. Don't judge. Just watch and note. What takes longer than it should? What requires multiple systems or approvals? Where do people develop workarounds? What makes them sigh with frustration?

While shadowing, calculate rough CRCs for common activities. You don't need precision. Just ask the following questions: How much time does this take? What value does it create? Who else benefits? What friction does it involve? You'll quickly see that most organizational activity has CRCs below 1:1. People are working hard to destroy value.

By Day 3, patterns will emerge. You'll see the same friction points across roles. The same wasteful practices. The same energy drains. Document these, but resist the urge to fix them yet. You're still in discovery mode.

Days 4 through 7, shift to quick wins. Find three atomic actions with CRCs above 5:1 that require no approval, no budget, and no system changes. These might be eliminating one recurring meeting that creates no decisions, creating a simple template that saves ten minutes per use, moving supplies closer to where they're used, or instituting a two-minute morning huddle that prevents confusion.

Implement these immediately. Don't announce them as part of a transformation. Just do them. Track the time saved, errors prevented, and energy created. You're not trying to transform the organization yet. You're proving to yourself that atomic actions work.

By Day 10, share your wins with one trusted colleague. Show them the math. If eliminating one meeting saved five hours per week across eight people, that's forty hours weekly. At loaded labor costs, that might be $50,000 annually. From killing one pointless meeting. Let that sink in.

Days 11–30: Building Momentum

Now you have proof. Time to build a coalition, but not the way traditional change management suggests. You don't need executive sponsorship yet. You need fellow practitioners who are tired of waste.

Find five people who showed enthusiasm for your quick wins. These aren't necessarily leaders or high performers. They're people who light up when they see a better way of working. Your atomic coalition should be diverse: include different departments, different tenure, and different perspectives. What unites them is frustration with friction.

Hold your first multiplication meeting on Day 15. Keep it under thirty minutes. Share the CRC concept simply. Show your three quick wins and their impact. Then ask the magic question: "What's one thing in your work that wastes time for no good reason?"

The floodgates will open. People have been carrying these frustrations for years. Document every suggestion but focus on finding five more atomic actions with CRCs above 5:1. Assign each coalition member one action to implement in their area. Set a date to reconvene in one week.

Days 16 through 22 are implementation week. Your coalition members are testing atomic actions in real time. Support them daily. Send a quick text: "How's your atomic action going?" When they hit obstacles, help them navigate. When they see success, celebrate immediately.

Your second multiplication meeting on Day 22 will have different energy. People will be excited to share the results. The person who simplified a form will report saving twenty minutes per day. The one who created a job aid will have prevented three errors. These are small wins, but the math is compelling. Calculate group CRC: total benefits divided by total effort. It will likely exceed 10:1.

Days 23 through 30, expand carefully. Each coalition member recruits one more person who's expressed interest. You now have 11 people implementing atomic actions. Create a simple tracking system, a shared document listing actions, implementers, and weekly benefits. No complex software. No formal process. Just visibility.

By Day 30, you'll have twenty to thirty implemented atomic actions generating measurable benefits. More importantly, you'll have momentum. People are asking to join. Success stories are spreading. The transformation has begun, and nobody's called it that yet.

Days 31–60: Systematic Implementation

Month two is when atomic transformation shifts from experiment to system. You need structure, but not bureaucracy. The balance is delicate.

Create an Atomic Action Board. This can be physical or digital, but it must be visible. Create three columns: Proposed, Testing, and Proven. Anyone can add to Proposed. Coalition members move items to Testing. After two weeks of measurable benefit, actions move to Proven. The board becomes a living display of progress.

Institute weekly Atomic Hours. Every Thursday from 2 to 3 PM, anyone interested gathers to share actions, calculate CRCs, and solve implementation challenges. Make it optional but valuable. People should leave with at least one new idea to try. Record sessions for those who can't attend.

Around Day 45, patterns emerge. You'll notice that certain types of atomic actions have consistently high CRCs: eliminating redundant approvals, creating visual management tools, simplifying communication protocols, and automating repetitive data entry. Create categories and look for multiplication opportunities within each.

Start measuring system indicators, not just individual actions. Track the answers these questions: How many people are proposing atomic actions? How quickly do actions move from proposed to proven? What's the average CRC of implemented actions? How many actions are spreading beyond their original area? These metrics reveal transformation health.

By Day 60, you'll face your first crisis. Someone, usually in middle management, will notice all this "unauthorized" improvement.

They'll raise concerns about control, standards, or process compliance. This is your moment of truth. Don't get defensive. Instead, show the data. Present the accumulated benefits. Calculate the annual impact if these improvements continue. Make it about results, not rebellion.

Days 61-90: Culture Shift

Days 61 to 90 is when atomic thinking either becomes embedded culture or withers into another failed initiative. The difference lies in how you handle scaling.

First, get official without getting bureaucratic. Schedule a presentation to senior leadership, but don't ask for permission or funding. Present atomic transformation as something that's already happening and generating proven returns. Show the math: accumulated benefits, projected annual impact, and zero cost of implementation. Position yourself as reporting success, not requesting support.

Create Atomic Champions in each department. These aren't appointed positions but recognized roles. Champions are simply the people generating the highest CRC improvements in their areas. Give them visibility, not authority. Let their success inspire others.

Launch the Multiplication Challenge. For one week, challenge everyone to find one atomic action with CRC above 10:1. Make it competitive but collaborative. Teams can combine actions for higher impact. Public recognition for winners, but everyone who participates gets acknowledged. You're building cultural permission to improve.

Address the antibodies directly. By now, you'll have identified the organizational antibodies: people who resist change, policies that prevent improvement, and systems that punish innovation. Don't fight them. Outflank them. Show how atomic actions make their jobs easier. Find atomic improvements that benefit them directly. Convert antibodies into allies by solving their problems.

Institute Failure Fridays. The last thirty minutes of each Friday, teams share atomic actions that didn't work and what they learned. This isn't blame; it's education. When people see that failed attempts are celebrated for their learning value, experimentation accelerates. Risk aversion melts away.

By Day 90, measure cultural indicators. What percentage of employees have implemented at least one atomic action? How many departments have active Atomic Champions? Are people using CRC language naturally? Do meetings start with atomic wins? Is "What's the CRC?" becoming a common question? These signals matter more than any financial metric.

Days 91-100: Sustaining the Chain Reaction

The final ten days determine whether atomic transformation becomes permanent or peters out. You need three things: systems that self-sustain, measurement that motivates, and leadership that multiplies.

Build the Atomic Operating System. Document the simple processes that make atomic transformation work: how to propose actions, how to test them, how to measure CRC, how to share learnings, and how to scale successes. Keep it to one page. Complexity kills adoption.

Create an Atomic Dashboard visible to everyone. Show cumulative benefits, number of active improvements, participation rate, average CRC, and trend lines. Update it weekly. When people see collective impact growing exponentially, contribution becomes contagious.

Transition from pioneer to platform. Your role shifts from implementing atomic actions to enabling others. Create tools, templates, and training that help anyone identify and implement high CRC improvements. Your new CRC isn't your personal actions but how many others you enable.

Prepare for the plateau. Around Day 100, you'll hit the first plateau. Easy wins are harvested. Enthusiasm normalizes. This is natural and temporary. Have your next level ready: cross-functional atomic projects, customer-facing improvements, and supplier collaboration opportunities. The second wave of atomic actions often has even higher CRCs.

Lock in leadership commitment. By now, senior leaders have seen the results. Get specific commitments: protecting Atomic Hours, including CRC in performance discussions, funding tools that enable atomic actions, and celebrating multiplication in company communications. You don't need a big program. You need sustained support.

The Atomic Transformation Checklist

Use this checklist to track your progress:

Days 1–10: Assessment

- Shadowed three different roles
- Identified major friction points

- Calculated rough CRCs for common activities
- Implemented three quick wins
- Documented time/cost savings
- Shared wins with one colleague

Days 11–30: Momentum

- Recruited five coalition members
- Held first multiplication meeting
- Implemented five more atomic actions
- Calculated group CRC exceeding 10:1
- Expanded coalition to eleven members
- Created simple tracking system

Days 31–60: Systematic

- Created Atomic Action Board
- Instituted weekly Atomic Hours
- Identified high-CRC patterns
- Measured system indicators
- Handled first resistance successfully
- Accumulated measurable benefits

Days 61–90: Culture

- Presented to senior leadership
- Identified Atomic Champions
- Ran Multiplication Challenge
- Converted key antibodies

- Instituted Failure Fridays
- Measured cultural adoption

Days 91–100: Sustaining

- Built Atomic Operating System
- Created visible dashboard
- Enabled others to lead
- Prepared for plateau
- Locked in leadership support
- Planned second wave

Measuring What Matters

Throughout your 100 days, track these essential metrics:

Participation Rate: Percentage of employees who've implemented at least one atomic action. Target: 20 percent by day 100.

Average CRC: Mean Chain Reaction Coefficient of all implemented actions. Target: 8:1 or higher.

Spread Velocity: How fast successful atomic actions spread to other areas. Target: three times multiplication within thirty days.

Sustained Adoption: Percentage of atomic actions still active 60 days after implementation. Target: 80 percent or higher.

Cultural Penetration: Frequency of atomic language in everyday conversation. Target: Heard daily by day 100.

System Health: Ratio of proposed to implemented actions. Target: 50 percent implementation rate.

ROI: Calculated annual benefit from all atomic actions divided by implementation cost. Target: 100:1 by day 100.

The Reality Check

Let me be clear about what you'll face. Days 15 through 25 will feel slow. You'll question whether this is worth the effort. Around Day 40, someone will actively resist. Near Day 70, you'll be tempted to formalize everything and kill the organic growth. By Day 85, you'll wonder if the culture is really changing or just pretending.

These are predictable challenges. Push through them. Every organization that has successfully implemented atomic transformation faced the same doubts. The difference between success and failure isn't the absence of obstacles. It's the persistence to continue implementing atomic actions despite them.

Your Next Action

You now have a complete roadmap for your first 100 days. But roadmaps don't create transformation. Action does. Here's your challenge:

Within twenty-four hours of reading this, identify one process in your work that has a CRC below 1:1. Calculate exactly how much time or resources it wastes. Find one small modification that would improve the CRC to above 5:1. Implement it tomorrow. Track the results for one week.

That's your first atomic action. It will take less than an hour to identify and implement, but it will prove to you that transformation isn't about grand strategies or executive mandates. It's about individuals like you deciding that waste is unacceptable and doing something about it.

One action becomes ten. Ten becomes a hundred. A hundred becomes a culture. A culture becomes transformation.

Your hundred days start now - ot with fanfare or announcement, but with one person, one action, one small improvement that starts a chain reaction.

The only question is this: what will your first atomic action be?

In 100 days, you'll look back and marvel at how far you've come, but you'll only get there by starting. Starting is what matters most.

Welcome to your atomic transformation. The countdown begins with your next decision.

Your Atomic Journey

Small actions. System thinking.
Exponential impact.

As I write these final words, I'm sitting in the same office where this journey began, surrounded by dissertation notes, Half Price Books finds on quantum physics, and countless stories of organizational transformation. What started as frustration with traditional productivity approaches has become a movement I see spreading across industries and continents.

This isn't my story; it's yours.

Your First Atomic Action

Don't try to transform everything at once. That's not atomic thinking. Instead, take a more methodical approach.

Tomorrow morning, identify one friction point that irritates you daily. Fix it. Calculate the CRC. Share the success.

That's it. One atomic action.

But here's what will happen. Others will notice. They'll want to share their own friction points. Solutions will multiply. Energy will build. The system will begin to transform.

I've seen it happen hundreds of times. It will happen for you too.

The Physics of Hope

Remember that small book on quantum theory that started this journey? It taught me that at the smallest scales, at the atomic level, enormous energy waits to be released. The same is true in organizations.

Every frustrated employee, every broken process, every customer complaint contains potential energy. Your job isn't to create that energy; it's to release it through better system design.

That's the physics of hope: We don't need more resources. We don't need heroic effort. We need to split the right atoms in the right way.

My Challenge to You

As you close this book, I challenge you to see differently. Look for atoms, not mountains. Search for multiplication, not addition.

Start small. Pick one atomic action this week - just one. Make it count.

Think systems. Ask not, "Who failed?" but "What about our system allowed this?"

Build chains. Connect your atomic action to others. Create reactions.

Share freely. Atomic knowledge multiplies when shared. Teach others.

Measure impact. Calculate your CRCs. What multiplies, thrives.

Stay human. Technology amplifies but humanity transforms.

The Atomic Promise

I promise you this: If you apply these principles, identify atomic actions, build supporting systems, create psychological safety, and measure what matters, you will transform your organization.

Not overnight. Not without setbacks. Not without resistance.

But inevitably. Relentlessly. Exponentially.

That's how atomic reactions work. One atom splits, releasing energy that splits others, creating a cascade of transformation that changes everything - an atomic impact!

Final Words

The atomic age of productivity isn't coming. It's here. In every organization, in every industry, in every country, people are discovering that small actions in the right systems create exponential impact.

The only question is this: will your organization be one of them?

The atoms are already there. The energy waits to be released. The willing workers are ready. All that's missing is you.

So, take that first atomic action. Calculate that first CRC. Build that first system improvement. Create that first chain reaction.

Once you start, once you see the multiplication in action, once you feel the energy of transformation, there's no going back.

Welcome to your atomic journey. Welcome to a world where small actions create massive impact. Welcome to the future of productivity.

It's atomic, and it starts now with you.

The power has always been there. You just needed to see it differently.

Now go split some atoms.

Glossary

Acceptability vs. Desirability Framework: A concept distinguishing between meeting minimum standards (acceptability) and achieving excellence that delights stakeholders (desirability).

Antifragility: Systems that don't just survive shocks but get stronger from them. Goes beyond resilience to actively benefit from volatility and stress.

Atomic Action: A small, specific intervention within an organizational system that creates chain reactions of productivity far exceeding its initial investment. Must have a CRC of 5:1 or greater.

Atomic Productivity Theory (APT): The framework proposing that small, strategically placed actions within well–designed systems can trigger exponential improvements in organizational productivity.

Chain Reaction Coefficient (CRC): A formula measuring the multiplicative impact of an action: (Primary Benefits + Secondary Benefits $\times$ 0.7) $\times$ Network Reach / (Direct Investment $\times$ 1.5). A CRC above 5:1 indicates atomic potential.

Common Cause Variation: Normal, expected variation within a stable system. Part of the system's natural behavior.

Control Rods: Organizational mechanisms that regulate the rate of change to prevent burnout while maintaining transformation momentum. Borrowed from nuclear reactor design.

Critical Mass: The threshold (typically 15–20% of actions being atomic) at which transformation becomes self–sustaining and multiplication effects take hold.

Distributed Intelligence: Organizational design where decision–making authority exists at multiple levels, particularly close to where work is performed.

Energy Multiplier: An atomic action that boosts human energy and motivation, creating positive chain reactions in engagement and performance.

Feedback Loops: System mechanisms that provide information about the impact of actions, enabling continuous adjustment and improvement.

Fission Action: An atomic action that naturally creates chain reactions. Has four key characteristics: natural multiplication, system alignment, low activation energy, and sustainable reaction.

Friction Eliminator: An atomic action that removes small barriers creating disproportionate delays or inefficiencies in systems.

Key Progress Indicators (KPgIs): Forward–looking metrics that show momentum and direction, as opposed to traditional KPIs that measure past performance.

Learning Velocity: The speed at which an organization learns from experience and adapts its practices. Critical for resilience and innovation.

Mental Models: The unconscious frameworks and assumptions that shape how we interpret and respond to situations.

Multiplication Effect: The phenomenon where atomic actions spread and evolve throughout an organization, creating exponential rather than linear improvement.

Network Reach: A component of the CRC formula measuring how many people or departments are impacted by an atomic action.

Non–Billable Overtime (NBOT): Time worked that cannot be charged to clients, used as an example of variation containing system information.

Open Loop Thinking: Linear thinking that doesn't account for feedback effects and system interactions.

Psychological Safety: An environment where people feel safe to take risks, make mistakes, ask questions, and challenge the status quo without fear of punishment or humiliation.

Quantum Productivity: Applying principles from quantum physics (superposition, entanglement, uncertainty) to organizational design and productivity.

Red Bead Experiment: Deming's demonstration showing how system design, not individual performance, determines most outcomes.

Special Cause Variation: Variation caused by factors outside the normal system operation. Indicates something has changed or gone wrong.

Statistical Process Control (SPC): Methods for understanding and managing variation in processes, distinguishing between common and special causes.

Superposition: In organizational terms, the ability to exist in multiple potential states until conditions demand crystallization into a specific form.

System Dynamics: The study of how systems behave over time, including feedback loops, delays, and non–linear effects.

System of Profound Knowledge (SoPK): Deming's framework encompassing appreciation for systems, knowledge of variation, theory of knowledge, and psychology.

System Steward: A leader who tends to system health and interactions rather than trying to control individual behaviors.

Technology Acceptance Model (TAM): Framework showing technology adoption depends on perceived usefulness and ease of use. In atomic systems, also includes perceived amplification and humanity.

Variation: Differences in performance or outcomes. In atomic thinking, variation is information about system health, not error to eliminate.

Willing Worker: Deming's concept that most employees come to work wanting to do a good job and contribute meaningfully. System failures, not worker failures, typically prevent excellence.

Bibliography

Primary Sources

Ackoff, Russell L. *Systems Thinking for Curious Managers*. Triarchy Press, 2010.

- Essential reading on systems thinking and the dangers of optimizing parts versus wholes.

Clear, James. *Atomic Habits: An Easy & Proven Way to Build Good Habits & Break Bad Ones*. Avery, 2018.

- While focused on personal habits rather than organizational productivity, provides useful context for understanding atomic–level change.

Deming, W. Edwards. *Out of the Crisis*. MIT Press, 1986.

- The foundational text on quality management and systems thinking, including the Red Bead Experiment and System of Profound Knowledge.

Deming, W. Edwards. *The New Economics for Industry, Government, Education*. MIT Press, 1994.

- Deming's later work expanding on systems thinking and transformation.

Edmondson, Amy. *The Fearless Organization: Creating Psychological Safety in the Workplace for Learning, Innovation, and Growth.* Wiley, 2019.

- The definitive work on psychological safety and its role in organizational performance.

Forrester, Jay W. *Industrial Dynamics.* MIT Press, 1961.

- Groundbreaking work on system dynamics and understanding complex organizational behaviors.

Gladwell, Malcolm. *The Tipping Point: How Little Things Can Make a Big Difference.* Little, Brown and Company, 2000.

- Exploration of how small changes can trigger large–scale transformation.

Rhodes, Richard. *The Making of the Atomic Bomb.* Simon & Schuster, 1986.

- Pulitzer Prize–winning account that inspired the atomic metaphor for organizational transformation.

Senge, Peter M. *The Fifth Discipline: The Art and Practice of the Learning Organization.* Doubleday, 1990.

- Classic work on systems thinking in organizations and creating learning cultures.

Sterman, John. *Business Dynamics: Systems Thinking and Modeling for a Complex World.* McGraw–Hill, 2000.

- Comprehensive textbook on system dynamics with practical applications.

Taleb, Nassim Nicholas. *Antifragile: Things That Gain from Disorder*. Random House, 2012.

- Introduction to the concept of antifragility and designing systems that benefit from volatility.

Secondary Sources

Buckingham, Marcus and Ashley Goodall. *Nine Lies About Work: A Freethinking Leader's Guide to the Real World*. Harvard Business Review Press, 2019.

- Challenges conventional management wisdom, including the reliability of performance ratings.

Davis, Fred D. "Perceived Usefulness, Perceived Ease of Use, and User Acceptance of Information Technology." *MIS Quarterly*, vol. 13, no. 3, 1989, pp. 319–340.

- Original paper introducing the Technology Acceptance Model.

Gibson, William. *Neuromancer*. Ace Books, 1984.

- Source of the quote "The future is already here – it's just not evenly distributed."

Shewhart, Walter A. *Economic Control of Quality of Manufactured Product*. Van Nostrand, 1931.

- Foundation work on statistical process control and understanding variation.

Smith, Adam. *An Inquiry into the Nature and Causes of the Wealth of Nations*. W. Strahan and T. Cadell, 1776.

- Classic economic text including the pin factory example of division of labor.

Taguchi, Genichi. *Introduction to Quality Engineering*. Asian Productivity Organization, 1986.

- Alternative approach to quality emphasizing robustness in the presence of variation.

Additional Resources

MIT System Dynamics Group: https://mitsloan.mit.edu/faculty/academic–groups/system–dynamics

- Ongoing research and resources on system dynamics.

The Deming Institute: https://deming.org

- Resources and continuing education on Deming's principles.

Connect with Dr. Gilbert Guzman

Thank you for reading *Atomic Impact*. The journey to transforming productivity through small, strategic actions doesn't end here – it begins with your first atomic action.

Continue Your Atomic Journey

Visit the Atomic Impact Hub www.atomicimpactbook.com
At my website, you'll find:

- Free downloadable tools and templates from the book
- Video tutorials on calculating CRC and identifying atomic actions
- Case study updates showing long–term transformation results
- Community forum to share your atomic actions and learn from others

Bring Atomic Impact to Your Organization

Dr. Guzman is available for:

Keynote Speaking Inspire your team with the power of atomic productivity. Dr. Guzman's energetic presentations blend storytelling, science, and practical application to create immediate impact.

Workshops and Training

- Half–day Atomic Action Identification Workshop
- Full–day Leadership for Atomic Impact Training
- Multi–day Atomic Transformation Boot Camp
- Custom programs designed for your specific industry and challenges

Consulting and Implementation Work directly with Dr. Guzman to implement atomic productivity in your organization. From initial assessment through full transformation, get expert guidance tailored to your unique system.

Get in Touch
Email: dr.guz@atomicimpactbook.com
LinkedIn: https://www.linkedin.com/in/gilguz/
Website: www.atomicimpactbook.com
Bulk Book Orders: For discounts on bulk orders for your team or organization, email orders@atomicimpactbook.com

Share Your Story

Have you implemented atomic actions in your organization? I'd love to hear about your experience. Email your atomic transformation story to: stories@atomicimpactbook.com

A Personal Note

When I started this journey, frustrated by traditional productivity approaches that demanded more resources and heroic efforts, I never imagined it would lead here. The stories of transformation I've witnessed – from midnight shift workers becoming company heroes to struggling stores becoming community anchors – continue to inspire me daily.

This book exists because of the willing workers trapped in broken systems who shared their stories with me. It thrives because of readers like you who see the potential for transformation through atomic actions.

Your organization has atomic potential waiting to be unleashed. The question isn't whether transformation is possible –it's when you'll take the first small step that creates a chain reaction of success.

What will your first atomic action be?

Remember: The power was always there. You just needed to see it differently.

–Dr. Gilbert A. Guzman
Fort Worth, Texas

P.S. If this book created value for you, please consider leaving a review on Amazon or Goodreads. Your review helps other leaders discover the power of atomic productivity. Thank you!

Appendix and Supporting Resources

CRC Calculation Worksheet

Chain Reaction Coefficient (CRC) = (Primary Benefits + Secondary Benefits × 0.7) × Network Reach / (Direct Investment × 1.5)

Step 1: Identify Your Atomic Action
Action Description: _____

Step 2: Calculate Primary Benefits
Direct time saved per occurrence: _____ minutes Frequency (daily/weekly/monthly): _____ People directly affected: _____ **Total Primary Benefit (in hours/month):** _____

Step 3: Calculate Secondary Benefits
☐ Improved morale/engagement (est. hours): _____ ☐ Error prevention (est. hours saved): _____ ☐ Knowledge transfer (est. learning time saved): _____ ☐ Customer satisfaction improvement (est. value): _____ ☐ Other: _____ (est. hours): _____ **Total Secondary Benefits:** _____ × 0.7 = _____

Step 4: Determine Network Reach

□ Just you = 1.0 □ Your immediate team (2-10 people) = 1.3 □ Your department (11-50 people) = 1.7 □ Multiple departments (51-200 people) = 2.2 □ Entire organization (200+ people) = 2.5 **Network Reach Score:** _____

Step 5: Calculate Direct Investment

Initial setup time: _____ hours Ongoing time per month: _____ hours Resources/tools needed: $_____ (convert to hours at avg. wage) **Total Investment:** _____ × 1.5 = _____

Step 6: Calculate Your CRC

(_____ + _____) × _____ / _____ = **CRC:** _____

Atomic Action Identification Template

Current State Analysis

1. What process/task are you examining?

2. Current pain points (check all that apply):
□ Takes too long
□ Requires multiple approvals
□ Involves switching between systems
□ Creates frequent errors
□ Causes frustration
□ Requires rework
□ Delays others
□ Wastes resources

3. Time currently required:

Per occurrence: _____ minutes

Frequency: _____ times per _____

Monthly total: _____ hours

Atomic Opportunity Discovery

4. What would the ideal state look like?

5. What's the smallest change that would create the biggest impact?

6. What barriers need to be removed?

☐ Unnecessary approvals

☐ Redundant data entry

☐ Physical distance

☐ System complexity

☐ Unclear instructions

☐ Missing tools

☐ Poor communication

☐ Other: _____

Implementation Planning

7. Can you implement this:

☐ Alone, immediately

☐ With team agreement

☐ With manager approval

☐ With IT support

☐ With budget approval

8. Pilot approach:

Start date: _____

Test duration: _____

Success metrics: _____

Who to involve: _____

Quick Reference Formulas

Chain Reaction Coefficient (CRC) = (Primary Benefits + Secondary Benefits × 0.7) × Network Reach / (Direct Investment × 1.5)

Technology Success Score = (Human Amplification × System Integration) / (Friction Added + Human Replacement)

Meta-Metric (MES) = (Decisions Improved + Behaviors Aligned + Innovations Sparked) / (Cost to Collect + Time to Analyze + Confusion Created)

System Health Indicators

- Change Absorption Capacity = Successful Changes / Attempted Changes
- Innovation Velocity = Time from Idea to Implementation
- Multiplication Rate = Actions Spread / Actions Implemented
- Cultural Penetration = Employees Using Atomic Actions / Total Employees

Key Thresholds

- 15-20% of actions atomic = Transformation tipping point

- CRC > 5:1 = True atomic action
- MES > 1 = Metric worth tracking
- Spread Rate > 3x in 30 days = Viral adoption

Book Club/Team Discussion Guide

Part I: Understanding Systems

Chapter 1-3 Discussion Questions:

1. How does viewing your organization as a system change how you think about problems?
2. Share an example of when you were a "willing worker" trapped in a bad system. What atomic action could have helped?
3. What variation in your workplace is currently seen as a problem but might actually be valuable information?

Team Exercise: Red Bead Simulation Have team members draw colored beans from a container. Despite identical effort, results will vary. Discuss what this reveals about performance management in your organization.

Part II: The Atomic Framework

Chapter 4-7 Discussion Questions:

4. Calculate the CRC for one common activity in your workplace. Are you surprised by the result?
5. What "fission actions" have you observed that seem small but create chain reactions?

6. Where does your organization need "control rods" to prevent change burnout?

Team Exercise: Atomic Action Hunt Spend 20 minutes individually identifying potential atomic actions. Share and calculate CRCs together. Which has the highest multiplication potential?

Part III: Implementation

Chapter 8-9 Discussion Questions:

7. How could your team meetings be redesigned to have a CRC above 5:1?
8. What prevents atomic actions from multiplying in your organization?

Team Exercise: Design a Guidance System Pick one atomic action and design a complete system to guide its implementation and spread.

Part IV: Measurement and Technology

Chapter 10-12 Discussion Questions:

9. What metrics does your organization track that have an MES below 1? Why do they persist?
10. Where is technology replacing rather than amplifying human capability in your workplace?

Team Exercise: Technology Audit List all technology tools your team uses. Score each on the Human-Technology Balance Formula. What needs to change?

Part V: The Psychology of Systems

Chapter 13-15 Discussion Questions:

11. Rate your team's psychological safety from 1-10. What would it take to increase it by 2 points?

12. How does leadership in your organization accidentally suppress atomic actions?

Team Exercise: Safety Building Each person shares one "failed" improvement attempt. Team celebrates the learning and brainstorms how to create safety for future attempts.

Part VI: Real-World Application

Chapter 16-19 Discussion Questions:

13. What crisis has your organization faced that revealed hidden atomic potential?

14. Create your own 30-day atomic transformation plan. What would you do first?

Team Exercise: First 100 Days Planning Using Chapter 19 as a guide, create a realistic 100-day plan for your team. Assign atomic actions and set check-in dates.

Facilitation Tips for Atomic Transformation Discussions
Before the Meeting:

- Have everyone calculate one CRC from their work
- Share the atomic action definition
- Set expectation: looking for multiplication, not just improvement

During Discussion:

- Start with wins: "What's working that we could multiply?"
- Keep energy high with 5-minute atomic action breaks
- Document all ideas, even "impossible" ones
- Calculate CRCs together for top ideas

After the Meeting:

- Assign one atomic action per person to test
- Set follow-up date within 7 days
- Create shared tracking document
- Celebrate every attempt, not just successes

Making It Stick:

- Institute "Atomic Fridays" for sharing wins and lessons
- Create visual CRC tracking boards
- Recognize monthly "Atomic Champions"
- Build atomic thinking into existing meetings, don't add new ones

Recommended Resources

Books That Shaped Atomic Productivity
Systems Thinking Foundations

- *Out of the Crisis* by W. Edwards Deming - The foundation of understanding systems and willing workers
- *The Fifth Discipline* by Peter Senge - Systems thinking for learning organizations
- *Systems Thinking for Curious Managers* by Russell Ackoff - Challenging traditional management assumptions

- *Business Dynamics* by John Sterman - Comprehensive guide to modeling complex systems

Quantum and Atomic Inspiration

- *The Making of the Atomic Bomb* by Richard Rhodes - Understanding chain reactions and exponential power
- *Quantum Theory: A Very Short Introduction* by John Polkinghorne - Accessible introduction to quantum concepts
- *Quantum Theory:* by Niels Bohr & Max Planck – The start of learning quantum theory
- *The Quantum World* by J.C. Polkinghorne - Deeper dive into quantum mechanics principles

Organizational Transformation

- *The Fearless Organization* by Amy Edmondson - Creating psychological safety
- *Atomic Habits* by James Clear - Personal habits perspective on small changes